XI International Symposium on Photon Echo and Coherent Spectroscopy (PECS 2017)

EPJ Web of Conferences Volume 161 (2017)

Svetlogorsk, Russia
16 - 21 September 2017

Editors:

Vitaly V. Samartsev
Alexey A. Kalachev
Andrei V. Naumov
Maxim G. Gladush
Kamil R. Karimullin

ISBN: 978-1-5108-5103-0

Printed from e-media with permission by:

Curran Associates, Inc.
57 Morehouse Lane
Red Hook, NY 12571

Some format issues inherent in the e-media version may also appear in this print version.

Printed by Curran Associates, Inc. (2017)

For additional information, please contact EDP Sciences – Web of Conferences at the address below.

EDP Sciences – Web of Conferences
17, Avenue du Hoggar
Parc d'Activité de Courtabœuf
BP 112
F-91944 Les Ulis Cedex A
France

Phone: +33 (0) 1 69 18 75 75
Fax: +33 (0) 1 69 28 84 91

contact-edps@webofconferences.org

Additional copies of this publication are available from:

Curran Associates, Inc.
57 Morehouse Lane
Red Hook, NY 12571 USA
Phone: 845-758-0400
Fax: 845-758-2633
Email: curran@proceedings.com
Web: www.proceedings.com

TABLE OF CONTENTS

edp sciences

EPJ Web of Conferences

Statement of Peer review

In submitting conference proceedings to EPJ Web of Conferences, I certify to the Publisher that I adhere to the **Policy on Publishing Integrity** of the journal in order to safeguard good scientific practice in publishing.

1. All articles have been subjected to peer review administered by the proceedings editors.
2. Reviews have been conducted by expert referees, who have been requested to provide unbiased and constructive comments aimed, whenever possible, at improving the work.
3. Proceedings editors have taken all reasonable steps to ensure the quality of the materials they publish and their decision to accept or reject a paper for publication has been based only on the merits of the work and the relevance to the journal.

Title, date and place of the conference

XI International Symposium on Photon Echo and Coherent Spectroscopy (PECS - 2017), September 16 – 21, 2017, Svetlogorsk, Russia

Proceedings editor(s):

Prof. Dr. Vitaly V. Samartsev
Prof. Dr. Alexey A. Kalachev
Prof. Dr. Andrei V. Naumov
Dr. Maxim G. Gladush
Dr. Kamil R. Karimullin

Date and editor's signature

October 15, 2017

17, avenue du Hoggar - PA de Courtabœuf – BP 112 - 91944 Les Ulis Cedex A (France)
Tél. : 33 (0)1 69 18 75 75 - Fax : 33(0)1 69 07 45 17 – www.edpsciences.org

EPJ Web of Conferences **161**, 00001 (2017) DOI: 10.1051/epjconf/201716100001
PECS-2017

XI International Symposium on Photon Echo and Coherent Spectroscopy (PECS - 2017): preface

K.R. Karimullin[1,2], A.A. Kalachev[3], A.V. Naumov[1,2], S.V. Sazonov[4], and V.V. Samartsev[3]

[1]Institute for Spectroscopy RAS, 108840 Moscow, Troitsk, Russia
[2]Moscow State Pedagogical University, 119991 Moscow, Russia
[3]E. K. Zavoisky Physical-Technical Institute, Kazan Sci. Center RAS, 420029 Kazan, Russia
[4]National Research Center "Kurchatov Institute", 123182 Moscow, Russia

The International Symposium on Photon Echo and Coherent Spectroscopy (PECS) is a regular quadrennial scientific meeting. Symposium originates in Kazan at the Physical-Technical Institute of the Kazan Branch of the USSR Academy of Sciences (now the E. K. Zavoisky Physical-Technical Institute), where in 1963 the phenomenon of the photon echo was theoretically predicted. The first PECS was held in 1973 as All-Union Conference shortly after the first in the USSR experiments on the photon echo in a ruby crystal. The second symposium was also held in Kazan in 1981. Later the Symposium was organized every four years: 1985 – Kharkov (Ukraine, USSR), 1989 – Kuibyshev (now Samara), 1993 – Volga-Laser-Tour'93, 1997 – Yoshkar-Ola, 2001 – Novgorod the Great, 2005 – Svetlogorsk, 2009 – Kazan, 2013 – Yoshkar-Ola.

PECS-2017 was organized in Svetlogorsk (Kaliningrad region, Russia) from 16th till 21th September 2017 by the E. K. Zavoisky Physical-Technical Institute (Kazan, Russia, www.kfti.knc.ru/eng) and the Institute for Spectroscopy of the Russian Academy of Sciences (Moscow-Troitsk, Russia, www.isan.troitsk.ru/en) on the base of the company "Center for Energy Efficiency" (Kaliningrad, Russia, http://www.cef39.com/en).

The PECS program covered such topics as photon echo and coherent transients, nonlinear interaction of radiation with matter, optical memory and information processing, spectroscopy of ultrafast processes, quantum optics and quantum technologies, atomic optics and nanooptics, spectroscopy and microscopy of single quantum emitters. Scientific program of PECS-2017 included 13 plenary lectures and 18 invited talks, 42 oral talks, 31 poster talks, authored in total by more than 200 scientists from 14 countries (Russia, Belarus, Ukraine, Great Britain, Italy, Germany, France, The Netherlands, Poland, Portugal, Romania, India, USA, and Republic of South Africa).

Within the framework of the Symposium, a competition for young scientists was held. The winner of the competition has received a recommendation from the Organizing

Committee to participate in the competition for Fresnel Prize for the Young Scientists from Quantum Electronics and Optics Division of the European Physical Society.

This issue of the EPJ Web of Conferences consists of the expanded version of abstracts of the selected talks presented at the Symposium. The issue is recommended for specialists who work in the field of nonlinear and coherent optics and spectroscopy, as well as for M.S. and Ph.D. students. The full program and details about the Symposium can be found on the web-site www.pecs.su). The next 12th International Symposium on Photon Echo and Coherent Spectroscopy will be organized by the E. K. Zavoisky Physical-Technical Institute and the Institute for Spectroscopy of the Russian Academy of Sciences in 2021.

Participants of the Symposium

Organizing committee of PECS-2017

The financial support is acknowledged from the Federal Agency for Scientific Organisations of Russia and from Russian Foundation of Basic Researches (project No 17-02-20495). Members of local Organizing Committee of PECS - 2017 are involved in leading scientific school of Russia NSh-7035.2016.2 "Spectroscopy of atoms, molecules and condensed matter".

Coherent quantum states in the laser-induced thin film nanocluster structures: optical and electrophysical properties

S. Arakelian*, A. Kucherik, S. Kutrovskaya, A. Osipov, A. Istratov, and I. Skryabin

Vladimir State University named after Alexander and Nikolay Stoletovs, Vladimir, 600000 Russia

Abstract. We studied in both theory and experiment the nanocluster structures of different types (due to topology and composition) taking into account the correlations in nanoparticle-ensemble by quantum states. The problem of high temperature superconductivity due to topological surface structures with localized states has been under our consideration. Dramatic enhancement of electroconductivity (in six orders!), observed in our experiments, may be discussed as a real tendency to high temperature superconductivity due to topological peculiarities of nanocluster system. Near-field photoluminescence spectra/optical response in hybrid nanosystems measured by us demonstrated a big efficiency in the nanoantenna respect.

1. A nanocluster system may be presented as a quantum object when it's characteristic size $a \sim \lambda_{dB}$, where λ_{dB} is the de Broglie wavelength. But the principal question is the value of effective mass of electrons in the nanocluster system. For typical semiconductors we can estimate λ_{dB} =25nm being an ordinary nanocluster size. Moreover, for inhomogeneous system the correlated states should be very strong for nanoclusters.

In our experiments we generated some different network nanostructures of both types/dimensions and compositions. Such procedures for both direct laser modification of the solid state surface and laser ablation from target in liquid to produce, first, the colloidal systems and, second, to deposit the nanoparticles from colloid on solid surface are described e.g. in [1] including the hybrid set-up.

In frame of nanocluster network as a Kronig-Penny quantum wells (cf. [1]) we have (in analogy with optics) a local Mie-resonance model for electron wave-functions in contrast with a slowly varying energetic structure being a Bragg-resonance model. In last case we have non-local interference effect, and can speak about correlation of electrons due to existing of spatial structure. We have to separate the two such topology based opportunities: multiple Raman-Nath diffraction (for thin periodic structure) and namely Bragg-diffraction (for bulk structure).

But for electrical conductivity the last case allow us to consider the opportunity to potentially enhance the electroconductivity in some fixed direction. Thus, we can control the directional diagram and indicatrices for the charge particle propagation. Selected directions, i.e. anisotropy, are equivalent to symmetry breaking being a fundamental basis for physical states [3]. In general, two mentioned above cases can be presented as the two-phase states on a specific phase-diagram transition between different spatial structures.

* Corresponding author: arak@vlsu.ru

2. The electrical transport properties in analogy with the tunnel and quantum correlated states are namely considered; the jump (hopping)/tunneling/thermal activation effects for electroconductivity have been observed in different our experiments (cf. [4,5]).

To analyze the size-depended effect we used a percolation theory by computer simulation procedure [6]. In addition, we take into account the fractal structures of the units, and we estimated the measured value of electroresistance R vs both nanoparticle size (a) and topology by fractal dimension (D) as well.

According to this we had a good fit for the resistivity by theory (R_{cal}) and experiment (R_{meas}) in a large range of the values (see Fig. 1) for gold(Au) nanostructures as an example. The principal result is a six orders difference (!) in values of R by topology variation only. The temperature dependences for bi-metallic (Au+Ag) film are also obtained by us (cf. [7]), and the activation energy for electroconductivity is estimated as 1.3 eV.

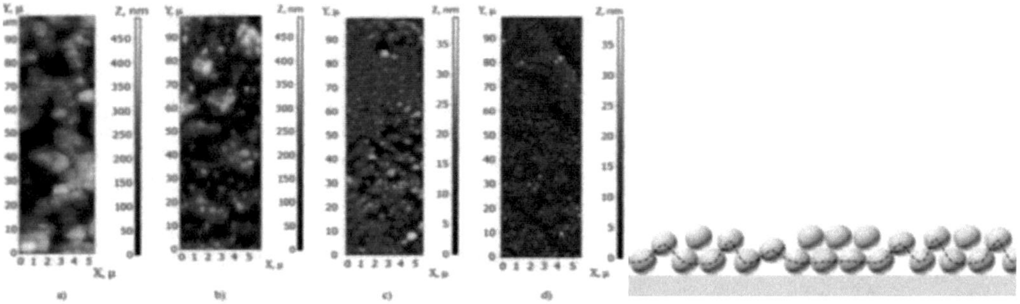

Fig. 1. Experimental and theoretical (by fractal approach) comparison of values for electroresistance R for Au-nanostructure (AFM-picture): a) $R_{meas}=5.4\cdot10^8$ Ohm, D=1.39, $R_{calc}=9.3\cdot10^8$ Ohm; b) $R_{meas} = 3.6\cdot10^7$ Ohm, D=1.57, $R_{calc}=5.5\cdot10^7$ Ohm; c) $R_{meas} = 5.4\cdot10^6$ Ohm, D=1.84, $R_{calc}=9.3\cdot10^6$ Ohm; d) $R_{meas} = 3.6\cdot10^2$ Ohm, D=1.93, $R_{calc}=5.5\cdot10^2$ Ohm, where R_{meas} – measurements; R_{calc} – calculations; D – fractal dimension. The size of the nanoobject: (a),(b) – 100 nm; (c),(d)–10nm. Optimal electron trajectory (in respect of min value of R), i.e. self-organization effect for the two layer system is shown (by dotted curve) in insert on the right.

3. In superconductivity problem (for e-fermions) the phenomenon can be presented as superfluidity (for bosons) when coupled electrons (Cooper-pairs) occur [3]. For cluster system some alternative mechanisms of electron coupling (not via standard phonon coupling) may exist. We discuss for the first time the several new opportunities for that.

4. Future activity in the field should be focused on the more detailed comparison of obtained experimental result for induced (by controllable way) nanocluster topological systems and concrete mechanism of its development under different conditions. The results give us perspective directions for possible applications of the phenomena in hybrid photonic devices (optics + electrophysics) of new generation, especially for optical signals enhancement in respect of nanoantenna (see e.g. [8]).

The work was partially carried out within the framework of RFBR projects 16-42-330461 p-a, 16-42-330531 p-a, 17-42-330928 p-a.

References

[1] S. Arakelian et al., *Introduction to the femtonanophotonics* (Moscow, Logos, 2015)

[2] V. Gartmakher, *Electrons in non-ordered medium* (Moscow, Fizmatlit, 2013).

[3] N.Bogolyubov, B.Sadovnikov, *Some problems in statistical mechanics* (Moscow, 1975)

[4] S. Arakelian et al., Opt. and Quant. Electronics **48(6)**, 342 (2016)

[5] S. Arakelian et al., J. Nanopart. Res. **18(6)**, 155 (2016)

[6] S. Arakelian et al., Opt. and Quant. Electronics **48(11)**, 505 (2016)

[7] A. Kavokin et al., Superlattices and Microstructures **109**, 1 (2017)

[8] S. Kutrovskaya S. Arakelian, A. Kucherik et al., Sci. Rep. **7**, 10284 (2017)

Study of ultrafast processes in matter by means of time-resolved electron diffraction and microscopy

Sergei Aseyev[1], *Boris* Mironov[1], *Sergei* Andreyev[1], *Victor* Kompanets[1], *Anatolii* Ischenko[2], *Igor'* Kochikov[3], *Oleg* Misochko[4], *Sergei* Chekalin[1,*], and *Evgenii* Ryabov[1]

[1]Institute of Spectroscopy, Russian Academy of Sciences, Troitsk, Moscow, 108840 Russia
[2]Moscow Technological University, 86 Vernadskogo ave., Moscow, 119571 Russia
[3]Moscow State University, 1 Leninskie Gory, Moscow, 119991 Russia
[4]Institute of Solid State Physics, Russian Academy of Sciences, Chernogolovka, 142432 Russia

Abstract. One of the most fundamental problems of modern natural science is the direct observation of atomic motions in the course of various processes. For this purpose, in the experiment it is necessary to provide high spatial-temporal resolution. The solution to this problem is achieved by using a pulsed electron beam of ultrashort duration to create a stroboscopic diffraction pattern in the method of time-resolved electron diffraction (TRED). Three types of experimental schemes have been developed at our lab. The experimental complex includes (i) 20-keV table-top apparatus for TRED, (ii) 75-keV ultrafast transmission electron microscope and (iii) lensless table-top device for femtosecond electron diffraction. The obtained experimental results are presented.

At present, the study of the structural dynamics of matter with a high spatially-temporal resolution is one of the key topics of modern researches. It is of undoubted interest for condensed matter physics, molecular and chemical physics, biophysics, and materials science [1–3]. Now ultrafast electron microscopy and electron diffraction are rapidly developing fields. These methods are based on probing coherent laser-induced processes in various materials using photoelectron bunches. This approach allows one to supplement the high temporal resolution provided by femtosecond lasers with the atomic spatial resolution, inherent in electron diffraction technique. As a result, one can make a video of the behavior of matter matched in 4D space – time continuum.

The first direct observation of the generation of coherent optical phonons in thin Sb film by TRED has been done using the experimental scheme, depicted in Fig. 1 [3]. The semimetal sample was excited by a femtosecond laser pulse and probed with a pulsed photoelectron beam. The vibrational modes of Sb lattice have two optical A_{1g} and E_g modes and two low-frequency A_{1u} and E_u acoustic modes. The fundamental of a femtosecond Ti:Sa laser was used as the pump beam. The pulsed photoelectron beam was formed under the irradiation of

* Corresponding author: chekalin@isan.troitsk.ru

the semitransparent silver photocathode by III harmonic of the same Ti:Sa laser and was used as the probe beam. In this manner, the optical and photoelectron pulses were strictly synchronized. The pump energy fluence on the sample was about 1.5 mJ/cm^2.

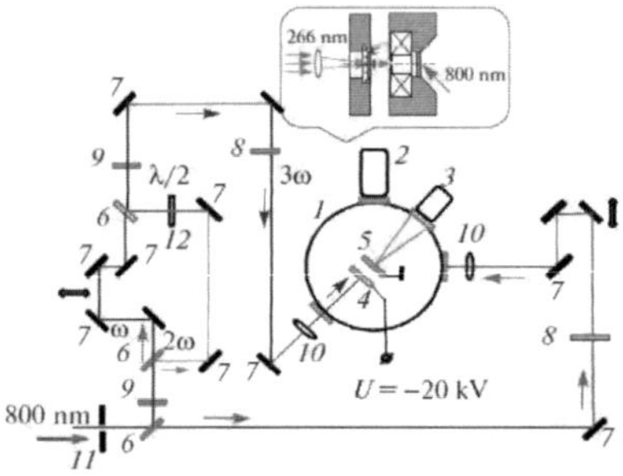

Fig. 1. Schematic of the setup for investigations by TRED with femtosecond resolution: (*1*) vacuum chamber; (*2*) turbomolecular pump; (*3*) position-sensitive detector based on MCP's; (*4*) silver photocathode; (*5*) anode + target; (*6*) laser beam dividers; (*7*) mirrors; (*8*) radiation attenuator; (*9*) converters of laser radiation into the second and third harmonics; (*10*) lenses; (*11*) diaphragm; (*12*) optical rotator. Inset shows the key parts with permanent magnetic lens. Gun-to-sample electron flight path is 12 mm.

In the experiment the variation of the intensity of the diffraction pattern caused by the optical excitation of the sample was studied as a function of the delay time between the pump and probe photoelectron pulses. We observed the sharp decay (with the characteristic time of about 300 fs, which corresponded to temporal resolution of the technique) in the diffraction signal after laser irradiation of the Sb film for positive temporal delay. Besides, the distinct oscillations with the period in sub-picosecond domain were also observed within about 20 ps. The Fourier analysis showed that the observed signal modulation was due to the presence of four distinct modes with frequencies of 1.1, 3.4, 4.6, and 6.4 THz (Fig. 2). Only three of these modes were observed earlier in the experiments on *optical probing* [4]. Knowing the attribution of the phonon modes in Sb from the literature, it is possible to associate the frequencies of 4.6 and 3.4 THz with the totally symmetric (A_{1g}) and twofold degenerate (E_g) optical phonons and 1.1 THz with the difference vibration A_{1g}–E_g. The frequency of 6.4 THz for antimony in the time domain was not observed earlier. It can be associated with the second harmonic E_g.

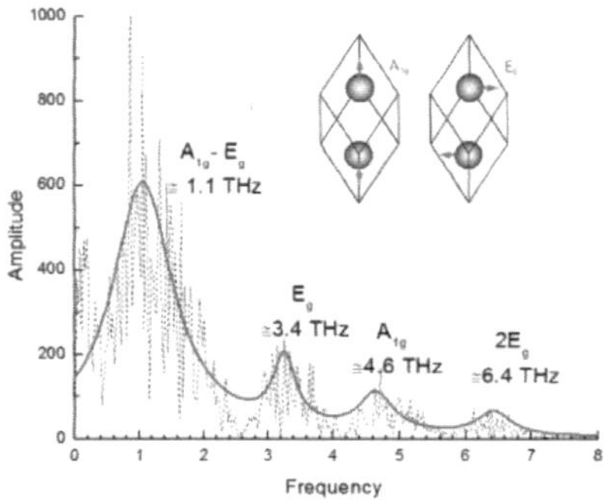

Fig. 2. The Fourier transformation of the normalized intensity of electron diffraction reflections in Sb film vs time after irradiation by femtosecond laser pulses. The signal modulation is due to the optical phonons illustrated in the inset.

Quite recently, at the Institute for Spectroscopy RAS an ultrafast transmission 75-keV photoelectron microscope was created [5]. The setup is based on commercial transmission electron microscope Hitachi H-300 and is intended for the researches of dynamic processes in the solids by excitation of a sample by femtosecond laser pulses and with probing emerging dynamics using ~ 7 ps pulsed photoelectron beam. The introduction of an adjustable optical delay between excitation and probing channels enables observation of ultrafast processes in matter. Our apparatus has been combined with 80-MHz femtosecond laser Mai-Tai, 2.5-MHz femtosecond fiber laser Antaus, or 1-kHz source of amplified femtosecond laser pulses with an energy of up to 4 mJ per laser pulse. In the first part of the experiments it was possible to visualize the interaction of picosecond 75-keV electron beam with charged cloud emitted from copper target under the action of focused femtosecond laser radiation. In the next part, the laser-induced plasma and associated processes in the vicinity of solid target will be investigated with high spatio-temporal resolution.

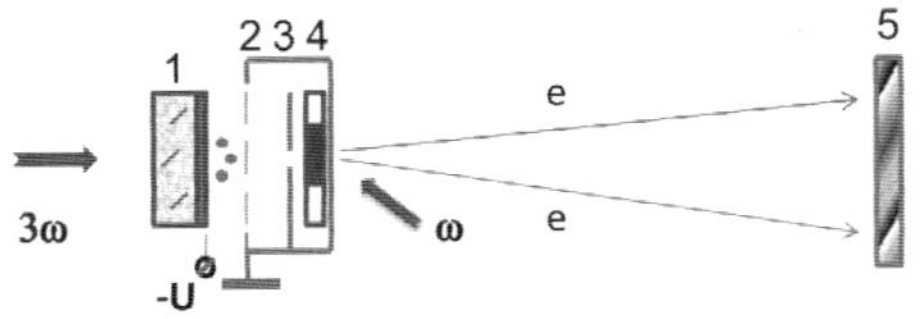

Fig. 3. Schematic diagram of lensless apparatus for TRED: (*1*) silver photocathode; (*2*) 0.5-mm-mesh; (*3*) 100-μ-diaphragm; (4) thin sample; (*5*) position-sensitive detector. U is accelerating voltage.

A distinguishing feature of the third instrument developed at our lab is the absence of collimating/focusing elements for the electron beam (Fig. 3). Since the planar silver cathode emits photoelectrons with minimal divergence and using a mesh anode further minimizes defocusing, the electron beam reaches the sample with nearly parallel momentum. The completely lensless design dramatically reduces the electron gun-to-sample electron flight path and therefore significantly improves the temporal resolution of TRED [6]. Comparing with the setup, depicted in Fig. 1, the lensless device (Fig. 3) is capable to work with different accelerating voltages. This opens wide opportunities for the experiments.

This study was supported by the Ministry of Education and Science of the Russian Federation (Project No. RFMEFI61316X0054).

References

[1] A.A. Ischenko, G.V. Girichev, Yu.I. Tarasov, *Electron Diffraction: Structure and Dynamics of Free Molecules and Condensed Matter* (Moscow, Fizmatlit, 2013)

[2] J.S. Baskin, A.H. Zewail, C.R. Phys. **15**, 176 (2014)

[3] B.N. Mironov, V.O. Kompanets, S.A. Aseyev, A.A. Ischenko, O.V. Misochko, S.V. Chekalin, E.A. Ryabov, JETP Lett. **103**, 531 (2016)

[4] K. Ishioka, M. Kitajima, O.V. Misochko, J. Appl. Phys. **103**, 123505 (2008)

[5] S.V. Andreyev, S.A. Aseyev, V.N. Bagratashvili, N.S. Vorob'ev, A.A. Ischenko, V.O. Kompanets, A.L. Malinovsky, B.N. Mironov, A.A. Timofeev, S.V. Chekalin, E.V. Shashkov, E.A. Ryabov, Quantum Electronics **47**(2), 116 (2017)

[6] D.S. Badali, R.Y.N. Gengler, R.J.D. Miller, Struct. Dynamics **3**, 034302 (2016)

EPJ Web of Conferences **161**, 01003 (2017) DOI: 10.1051/epjconf/201716101003
PECS-2017

On the criteria for strong and weak polarization responses of ordered objects and systems

Yakov Fofanov[1,*], *Igor* Sokolov[2], *Ivan* Pleshakov[2,3], *Vasiliy* Vetrov[4], *Andrey* Prokofiev[2,3], *Aleksey* Kuraptsev[3], and *Efim* Bibik[5]

[1]Institute for Analytical Instrumentation RAS, 190103, St. Petersburg, Russia
[2]Ioffe Physical Technical Institute RAS, 194021, St. Petersburg, Russia
[3]Peter the Great St. Petersburg Polytechnic University, 195251, St. Petersburg, Russia
[4]S. I. Vavilov State Optical Institute, 199034, St. Petersburg, Russia
[5]St. Petersburg State Technological Institute, 190013, St. Petersburg, Russia

Abstract. The fundamental system of criteria for strong and weak polarization responses of ordered objects and systems is considered. It is shown that the weak responses have important properties of linearity (proportionality) and additivity. As an example, the responses of magnetically ordered materials, optical crystals and ensembles of magnetic nanoparticles are considered. When scanning laser probing of the modified leucosapphire crystals, oscillations of weak polarization responses in laser radiation scattered on texture of crystals are recorded. Investigations of magnetic fluids polarization responses have shown their high sensitivity to introduction of coagulating impurities and to processes of agglomeration of magnetic nanoparticles.

The prospects for the further development and application of high sensitive methods for laser polarization-optical probing of matter may be limited in a number of cases by the lack of sufficiently complete ideas about their capabilities. Here some aspects of quantitative analysis of polarization responses for objects and systems with small optical anisotropy are considered [1,2].

A natural measure of the smallness of the observed optical anisotropy Δ can be the inequality $\Delta << \Delta_{LSL} \sim 1$ rad (here LSL – linearity scale limit). It is possible to call the studied objects (samples) meeting this criterion the samples with "weak anisotropy" or "weak samples", and the polarization responses corresponding to them (signals in the optoelectronic channel) - "weak responses" (or "weak signals"). In this case the observed responses have important properties of linearity (proportionality) and additivity.

Two other criteria take into account the photon (shot) noise of probing light, that considered as an instrument for researches (shot noise limit - SNL), and thermal noise of the registration system (thermal noise limit – TNL). To ensure that these noises do not affect the results of observations, polarization responses must be strong in according to both criteria: $\Delta >> \Delta_{SNL}$ and $\Delta >> \Delta_{TNL}$. In accordance with these criteria, the region $\Delta << \Delta_{SNL}$ may naturally be called the region of superweak (or superlow) signals – SWS [3-6].

* Corresponding author: yakinvest@yandex.ru

The additivity of weak polarization responses allows the use of a scalar approximation (scalar model), which does not take into account the polarization of light. However, this can lead to incorrect results and in the general case only the vector model should be used [7].

Thus, when the objects under study are probed by laser radiation with a strong polarization modulation, there is a very wide (about to seven orders of magnitude) and a very convenient range of measured responses from about $\Delta_{SNL} \sim 10^{-4}$ ang. min. up to values of the order of $\Delta_{LSL} \sim 10^{3}$ ang. min. The examples of objects which weak polarization responses characterize the processes of structure formation, ordering, agglomeration, etc. happening in them are further considered.

It was found that polarization magneto-optical responses of model magnetic fluid (magnetite nanoparticles in kerosene) are proportional to volume concentration of solid-phase in the range which is about four orders (10^{-2} - 10^{-6}), [8]. Fig. 1 illustrates the qualitative change of the polarization response of the investigated fluid in the case of addition of the coagulating impurity (an excess amount of surfactant).

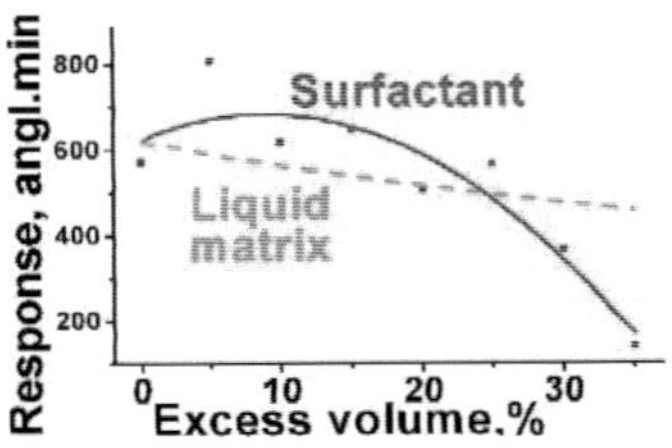

Fig. 1. Excess of surfactant.

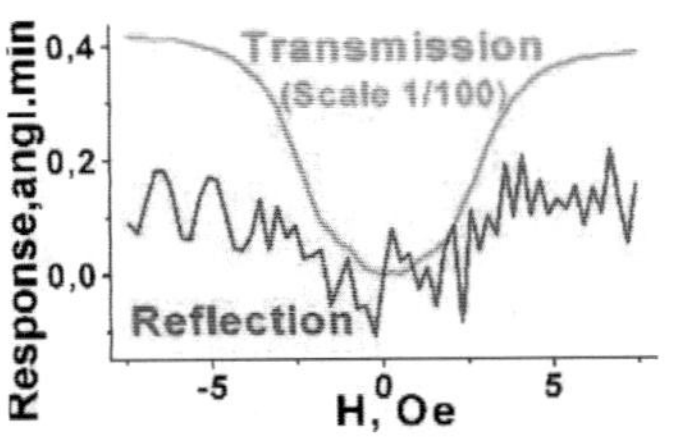

Fig. 2. Response of iron borate.

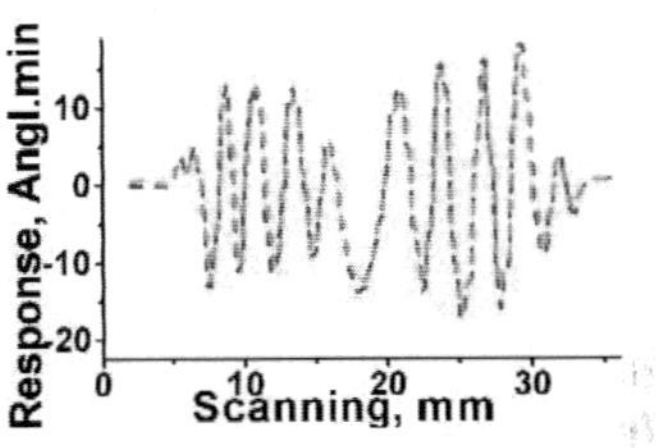

Fig. 3. Scattering in polar. lens

Polarization magneto-optical responses of an iron borate crystal at observation on transmitted and reflected light are presented in Fig. 2. It is seen that in the region of unsaturated magnetic fields the characteristic quadratic dependence of observed weak responses on the field is observed (Kotton-Mouton effect) [9].

Fig. 3 shows the oscillations of a weak polarization response in laser radiation scattered by the polarization lens material (leucosapphire). These oscillations indicate existence of textural heterogeneity and the formation of strong polarization structure due to the modification (bending) of the original crystal of leucosapphire [10].

This study was carried out with the financial support of the Russian Foundation for Basic Research (Grant No. 15-02-08703 a).

References

[1] Ya.A. Fofanov, Proceedings SPIE **1811**, 413 (1992)

[2] I.M. Sokolov, Ya.A. Fofanov, J. Opt. Soc. Am. A **12**, 1579 (1995)

[3] Ya.A. Fofanov, *Advances in Optoelectronics Research* (Nova Science Publishers, Inc., USA, 2014), 75–114

[4] Y.A. Fofanov, Proceedings SPIE **7993**, 79930O (2011)

[5] A.V. Belinsky, D.V. Volkov, A.V. Dmitriev, M.Kh. Shulman, JETP **117**, 771 (2013)

[6] A.V. Belinsky, M.Kh. Shulman, Physics-Uspekhi **57**, 1022 (2014)

[7] Ya.A. Fofanov, A.S. Kuraptsev, I.M. Sokolov, M.D. Havey, Phys. Rev. A **84**, 053811 (2011)

[8] Ya.A. Fofanov et al., Theor. Phys. Lett. **42**, 1054 (2016)

[9] Ya.A. Fofanov, I.V. Pleshakov, Yu.I. Kuz'min, J. Opt. Technol. **80**, 64 (2013)

[10] V.N. Vetrov, B.A. Ignatenkov, Crystallography Reports **58**, 474 (2013)

Solid-state Raman quantum memory in whispering gallery mode resonators: signal-to-noise ratio

Alexander Berezhnoi[1], and *Alexey* Kalachev[1,2,*]

[1]Kazan Federal University, 18 Kremlevskaya str., Kazan, 420008 Russia
[2]Kazan E. K. Zavoisky Physical-Technical Institute, 10/7 Sibirsky tract str., Kazan, 420029 Russia

Abstract. The possibility of implementation of optical quantum memory via off-resonant Raman absorption and emission of single-photon pulses in rare-earth-ion-doped crystals is theoretically analysed taking into account signal-to-noise ratio at the output of the memory device. The crystal $^{143}Nd^{3+}:Y^{7}LiF_4$ is considered as an example. It is shown that the signal-to-noise ratio can exceed unity for single-photon input pulses provided that storage and retrieval of them is performed in the doped crystals forming a microcavity such as whispering gallery mode resonator.

1 Introduction

Quantum memories are of crucial importance for developing quantum information technologies and form a platform for building scalable linear optical quantum computers, realizing long-distance quantum communications, etc. (see [1, 2] for a recent review). One of the most commonly discussed materials are rare-earth-ion-doped solids [3], in which the phase relaxation time at cryogenic temperatures may be as long as several hours [4]. Among them, isotopically pure crystals are of particular interest. They can demonstrate very small inhomogeneous broadening of optical transitions, reaching tens of MHz, which proves to be smaller than the hyperfine splitting of the energy levels of impurity ions. As a result, these crystals are promising candidates for implementing memory protocols based on off-resonant Raman interaction [5-7]. In the present work, we analyse the possibility of implementing optical quantum memory via off-resonant Raman absorption and emission of single-photon pulses in such crystals placed in a cavity and achieving large signal-to-noise ratio at the output of the memory device.

2 Basic results

To consider cavity-assisted quantum memory, we take the advantage of a theoretical model developed in [8]. The model is applied for analysis of the signal-to-noise ratio which

* Corresponding author: a.a.kalachev@mail.ru

can be obtained in the isotopically purified crystal $^{143}Nd^{3+}:Y^{7}LiF_4$, which has been used recently in experiments [9, 10] for implementing atomic-frequency-comb-based protocols of quantum memory. It is shown that the signal-to-noise ratio can exceed unity for single-photon input pulses provided that storage and retrieval of them is performed in the doped crystals forming a whispering gallery mode (WGM) resonator.

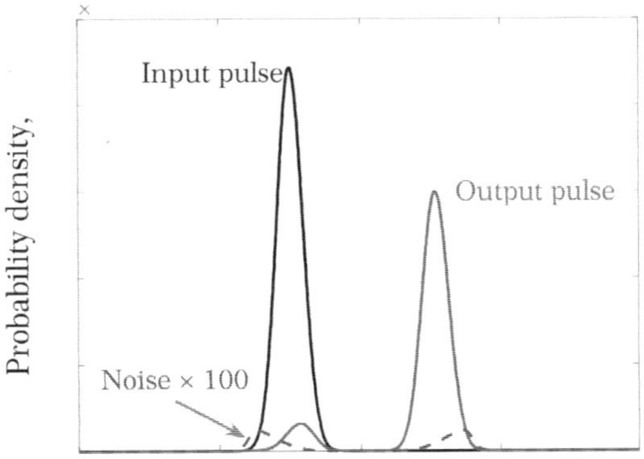

Fig. 1. Illustrating storage and retrieval of single-photon Gaussian pulses in $^{143}Nd^{3+}:Y^{7}LiF_4$ crystal.

To be more precise, we consider a WGM resonator of 200 μm in diameter with Q-factor of ~ 10^8. The latter corresponds to the cavity finesse of ~ 10^5, which is a typical value for such a type of resonators. In addition, we consider an ideal Λ-scheme corresponding to ZEFOZ spin transition that can be formed in this crystal under application of a weak magnetic field [8]. Then, according the numerical simulation, we can obtain less than 1% of noise at the output of the memory, which strongly depends on the number of impurity atoms. In particular, taking the dipole moment of the optical transitions 0.003 D and the maximum value of the control field Rabi frequency 8.4 MHz, we obtain 0.5% (0.08%) of noise for $1.4 \cdot 10^5$ ($7 \cdot 10^4$) ions. Fig. 1 displays the result of the simulation for Gaussian pulses of 425 ns duration.

This work is supported by the Russian Science Foundation (grant No. 14-12-00806).

References

[1] F. Bussières, N. Sangouard, M. Afzelius et al., J. Mod. Opt. **60**, 1519 (2013)
[2] K. Heshami, D.G. England, P.C. Humphreys et al., J. Mod. Opt. **63**, S42 (2016)
[3] C.W. Thiel, T. Böttger, R.L. Cone, J. Lumin. **131**, 353 (2011)
[4] M. Zhong, M.P. Hedges, R.L. Ahlefeldt et al., Nature **517**, 177 (2015)
[5] X. Zhang, A. Kalachev, O. Kocharovskaya, Phys. Rev. A **87**, 013811 (2013)
[6] A. Kalachev, O. Kocharovskaya, Phys. Rev. A **88**, 033846 (2013)
[7] X. Zhang, A. Kalachev, O. Kocharovskaya, Phys. Rev. A **90**, 052322 (2014)
[8] A. D. Berezhnoi, A.A. Kalachev, Quant. Electron. **47**, 790 (2017)
[9] R.A. Akhmedzhanov et al., Las. Phys. Lett. **13**, 015202 (2016)
[10] R.A. Akhmedzhanov et al., Las. Phys. Lett. **13**, 115203 (2016)

EPJ Web of Conferences **161**, 01005 (2017) DOI: 10.1051/epjconf/201716101005
PECS-2017

Dispersive shock waves in nonlinear and atomic optics

Anatoly Kamchatnov*

Institute of Spectroscopy, Russian Academy of Sciences, 108840, Moscow, Troitsk, Russia

Abstract. A brief review is given of dispersive shock waves observed in nonlinear optics and dynamics of Bose-Einstein condensates. The theory of dispersive shock waves is developed on the basis of Whitham modulation theory for various situations taking place in these two fields. In particular, the full classification is established for types of wave structures evolving from initial discontinuities for propagation of long light pulses in fibers with account of steepening effect and for dynamics of the polarization mode in two-component Bose-Einstein condensates.

Dispersive shock waves (DSWs) are universal phenomena of nonlinear physics. They were observed in water waves, plasma, solid state physics, nonlinear optics and dynamics of Bose-Einstein condensates (BECs). As examples of DSWs in the last two fields of research, one can mention observation of DSWs in BEC under action of a repulsive potential [1] and formation of shocks in evolution of light pulses in fibers [2]. In this report, we give a brief review of phenomena that have been recently observed and provide introduction into theoretical methods of their description. The main attention is paid to situations when the standard approach based on the Whitham theory for the NLS (Gross-Pitaevskii) equation is not applicable. For example, long distance propagation of light pulses in fibers can depend essentially on effects of retardation of the medium response on the electromagnetic field. Dynamics of polarization mode in BECs is even more complicated. In both these cases the nonlinear dispersionless velocities are not monotonous functions of the pulse parameters that leads to appearance of new types of wave structures evolving after wave breaking of the pulse. Such wave structures include kinks, trigonometric shocks, combined shocks, and their appearance makes the problem of classification of possible wave patterns much more difficult than in standard and well-known situations described by the NLS equation.

In this talk, we consider approach to the formulated above problem that is based on the developed earlier Whitham theory for the DNLS equation and Landau-Lifshitz equation for magnetics with easy-plane anisotropy. Generally speaking, the suggested method is applicable to nonlinear wave equations that are completely integrable in framework of the Ablowitz-Kaup-Newell-Segur scheme when the inverse scattering transform can be developed. In this case, the reduced finite-gap integration method yields the periodic solutions parameterized by the zeroes of the resolvent $R(\nu)$ of the polynomial $P(\lambda)$ that defines the solution whereas the zeroes of $P(\lambda)$ play the role of the Riemann invariants of the

* Corresponding author: kamch@isan.troitsk.ru

corresponding Whitham modulation equations [3]. The specific feature of the new approach is that the mapping of Riemann invariants to resolvent's zeroes is multi-valued and, as a result, one solution of the Whitham equations maps to two different DSW structures [4].

We apply the developed method to consideration of the so-called Riemann problem for evolution of initial discontinuities. The full classification of possible wave structures is established for propagation of long pulses in fibers with account of the steepening effect [5] and for polarization waves in BEC [6,7]. Example of a typical wave pattern is shown in Fig. 1. We present a simple algorithm which permits one to predict parameters of such wave patterns for given initial data at the discontinuity.

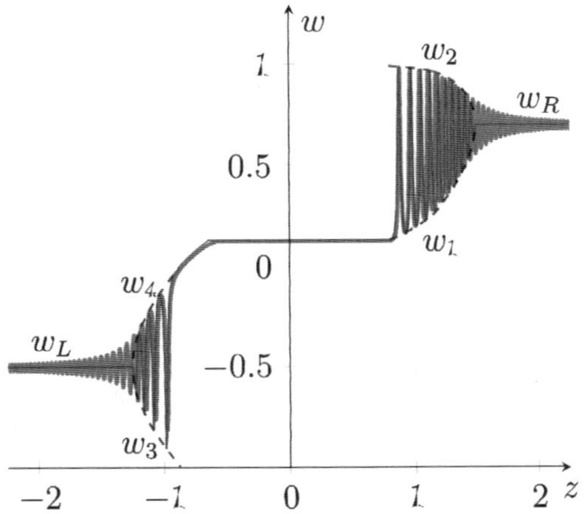

Fig. 1. Example of a wave structure evolving from an initial discontinuity in the polarization mode of two-component BEC dynamics, w is the polarization parameter; z is a coordinate along a trap. Analytical and numerical solutions are shown by blue and red lines, correspondingly.

This work was supported the Program "Topical problems of low temperature physics" of Russian Academy of Sciences.

References

[1] M.A. Hoefer, M.J. Ablowitz, I. Coddington et al., Phys. Rev. A **74**, 023623 (2006)

[2] G. Xu, M. Conforti, A. Kudlinski et al., Phys. Rev. Lett. **118**, 254101 (2017).

[3] A.M. Kamchatnov, *Nonlinear Periodic Waves and Their Modulations,* (World Scientific, Singapore, 2000).

[4] M. Kamchatnov, Y.-H. Kuo, T.-C. Lin, T.-L. Horng, S.-C. Gou, R. Clift, G.A. El, R.H.J. Grimshaw, Phys. Rev. E **86**, 036605 (2012).

[5] S.K. Ivanov, A.M. Kamchatnov, Riemann problem for the photon fluid: self-steepening effects (in preparation).

[6] T. Congy, A.M. Kamchatnov, N. Pavloff, Sci. Post. Phys. **1**, 006 (2016).

[7] S.K. Ivanov, A.M. Kamchatnov, T. Congy, N. Pavloff, M.A. Hoefer, E. Iacocca, Solution of the Riemann problem for polarization waves in a two-component Bose-Einstein condensate (in preparation).

EPJ Web of Conferences **161**, 01006 (2017) DOI: 10.1051/epjconf/201716101006
PECS-2017

Light bullets in transparent dielectrics

Valerii Kandidov[1,*], *Sergey* Chekalin[2], *Victor Kompanets*[2], and *Alexander* Dormidonov[1]

[1]Moscow Lomonosov State University, Physics Department and International Laser Center, 119991 Moscow, Russia
[2]Institute of Spectroscopy RAS, 108840 Moscow, Troitsk, Russia

Abstract. The state of research of the light bullets (LB) formation in the process of femtosecond laser pulse filamentation is presented. LB is a near single-cycle wave packet that is formed in the result of the light field self-organization in a nonlinear dispersive medium under matched spatiotemporal radiation self-compression in the regime of anomalous group-velocity dispersion (GVD). The formation of each LB is accompanied by the generation of a discrete portion of supercontinuum (SC) in the anti-Stokes region. LB is a short-lived robust object with parameters determined by fundamental properties of the medium and the laser pulse central wavelength.

A LB is an extremely compressed optical wave packet that is formed in the process of femtosecond laser pulse filamentation in transparent dielectrics in the regime of anomalous GVD. In the contrast to the compression of a light beam during self-focusing and to the shortening of an optical pulse duration during phase self-modulation, the formation of a LB with a high energy density occurs as a result of a joint and matched compression of laser radiation both in space and in time with self-action in a nonlinear dispersive medium [1]. Typical LB duration is 1 – 2 optical oscillations, diameter is about few wavelengths, and peak intensity exceeds 10^{14} W/cm^2.

A strong light field in LB leads to the ionization and generation of a low temperature laser plasma, the defocusing in which leads to sharp intensity decreasing on the LB trailing edge. This, in turn, is followed by short wavelength superbroadening of SC spectrum and formation of the narrow isolated anti-Stokes wing (ASW). ASW's bandwidth becomes narrower and its anti-Stokes shift increases with increasing of the initial pulse central wavelength (Fig. 1) [2,3]. The formation of a narrow ASW and a broad minimum in the SC spectrum, which separates ASW from the initial pulse central wavelength, is the result of constructive and, correspondingly, destructive interference of broadband SC radiation generated by a moving LB in a dispersive medium. When a laser pulse peak power is much greater than the critical power for self-focusing a sequence of LBs is formed in the medium. Each LB of the sequence ejects the same portion of energy in the anti-Stokes area of SC [4].

Near single-cycle LB shows a regular 'breathing' in filament that is a cyclic transformation of the light field amplitude and LB's diameter caused by the phase shift between the carrier wave and the wave packet envelope in a dispersive medium. It is shown

* Corresponding author: kandidov@physics.msu.ru

DOI: 10.1051/epjconf/201716101006

that LB 'breathing', which was revealed by the observation of strictly periodic color centres structures created in isotropic LiF, directly reproduces the influence of extremely compressed wave packet absolute carrier wave phase on nonlinear-optical interactions with medium [5,6]. In the regime of multipulse filamentation induced color centres form a waveguide, during propagation in which the LB's lifetime increases significantly and the frequency-angular spectrum of SC changes qualitatively.

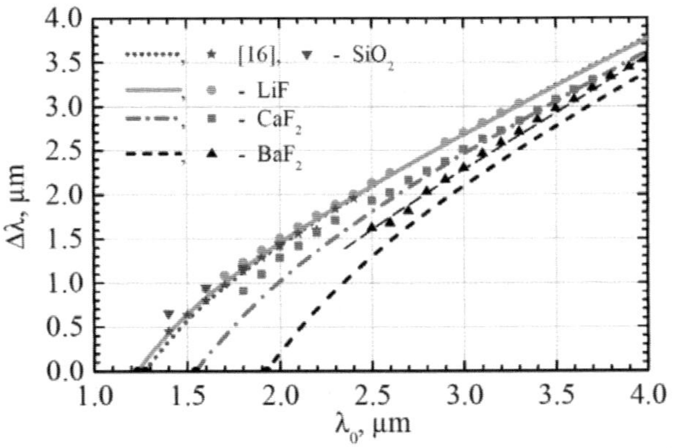

Fig. 1. Shift $\Delta\lambda$ of the wavelength of the SC maximum in the anti-Stokes region from the central wavelength λ_0 of the laser pulse in process of filamentation in regime of anomalous GVD in various materials.

LB is a short-lived robust formation that doesn't depend on spatial and temporal distribution of the light field amplitude and phase in the laser pulse and its parameters are determined by fundamental properties of matter. The LB and LBs sequence filamentation dynamics in transparent media under anomalous GVD have the same general scenario for condensed matter and gas. LB is formed both in a spectrally limited and chirped pulse, both in a collimated beam and in a beam focused by a lens or an axicon [4]. The optimum mode of LB formation is achieved with a matched compression of radiation in both time and space, which is feasible at equal diffraction and dispersion lengths. In this case, a robust LB is formed, the parameters and the spectrum of which don't depend on the energy of initial laser pulse.

References

[1] S.V. Chekalin, V.P. Kandidov, Physics-Uspekhi **56**, 123 (2013)
[2] A.E. Dormidonov, V.O. Kompanets, S.V. Chekalin, V.P. Kandidov, Opt. Express **23**, 29202 (2015)
[3] E.O. Smetanina, V.O. Kompanets, S.V. Chekalin, A.E. Dormidonov, V.P. Kandidov, Opt. Lett. **38**, 16 (2013)
[4] S.V. Chekalin, A.E. Dokukina, A.E. Dormidonov, V.O. Kompanets, E.O. Smetanina, V.P. Kandidov, J. Phys. B **48**, 094008 (2015)
[5] S.V. Chekalin, V.O. Kompanets, A.E. Dormidonov, V.P. Kandidov, Quant. Electron. **47**, 259 (2017)
[6] S.V. Chekalin, V.O. Kompanets, A.V. Kuznetsov, A.E. Dormidonov, V.P. Kandidov, Las. Phys. Lett. **13**, 065401 (2016)

Spatial phase light modulator based on the ferroelectric liquid-crystals: possibility of the device construction

Svetlana Kotova[1*], *Aleksandra* Mayorova[1], *Evgeny* Pozhidaev[2], and *Sergey* Samagin[1]

[1]Lebedev Physical Institute RAS, Samara Branch, 221 Novo-Sadovaya Str., Samara, 443011 Russia
[2]Lebedev Physical Institute RAS, 53 Leninskiy Prospekt, Moscow, 119991 Russia

Abstract. The work investigates the possibility for building of zonal spatial phase light modulators employing the quadratic electro-optic effect in ferroelectric liquid-crystals and capable of providing modulation frequencies up to 1 kHz. The study is made through the use of numerical simulation methods and experimental modulation characteristics

One of important challenges of modern optics is to increase the speed of control of the spatial characteristics of light fields. This task is relevant both to traditional areas of optics, including adaptive optics and holographic storage devices, and for such intensively developing fields of science and technology as holographic displays, optical manipulation of microscopic and nano-scale objects; quantum cryptography; quantum computation. As a rule for this the devices capable of pure phase modulation of light are needed. The traditional way of implementing a pure phase spatial modulation consists in the use of orientation effects in nematic liquid crystals (LC), for which characteristic frequencies are several tens of Hz. Such performance (response speed) restricts the possible scope of application for spatial light modulators (SLM) based on nematic LC. It is possible to increase in performance due to the orientation effects in ferroelectric liquid crystals (FLC) with sub-wavelength pitch of the spiral structure, in particular, orientational Kerr effect [1,2].

Under this work the possibility of constructing spatial modulators of different types based on the cell with a ferroelectric liquid crystal with sub-wavelength pitch of the spiral and with the helicoid axis orientated along the substrates, is analyzed by the numerical simulation method.

As soon as electric voltage is applied to the standard FLC cell with planar orientation of the main axis of the helix, then the change of the optical axis of the FLC layer direction and refractive indexes occurs. The change of the position of the optical axis of FLC in the plane perpendicular to the incident light wave direction causes the change in the state of polarization of the transmitted light. In this paper we studied the possibility of formation of light fields with the required spatial parameters in such conditions. Special attention was paid to the formation of light fields in the form of specified curves possessing orbital angular momentum.

* Corresponding author: kotova@fian.smr.ru

Electro-optical characteristics of FLC 587 were measured. Approximation of the experimental data allowed us to obtain required for the simulation of the SLM modulator based on FLC, dependencies of refractive indices, birefringence, effective phase delay on the value of the electric field applied to the layer.

Multi-pixel modulators with a square configuration of pixels and sector type modulators operating both in transmission and reflection were considered. Calculations of the light fields were made for the following scheme. The SLM was lightened with the linearly polarized light beam of homogeneous intensity. The value of the voltage fed to each pixel was determined in accordance with the required value of the effective phase delay which should be formed for the transmitted wave. The light transmitted or reflected by the modulator (depending on the modulator type) passed through the polarizer with the plane of transmission coinciding with the initial direction of polarization and was focused by the lens.

The light fields with nonzero orbital angular momentum and the distribution of intensity in the form of ring, curve (Archimedes spiral) and a closed curve (triangle) were obtained. Figure 1 shows an example of the formation of axially symmetric fields with the topological charge of 3 by means of a modulator of the sector type (12 sectors, the FLC layer thickness of 50 μm, diameter 1 cm). For this case and the other ones it was found that the total energy losses reached approximately 11%, the average deviation of the intensity did not exceed 5%.

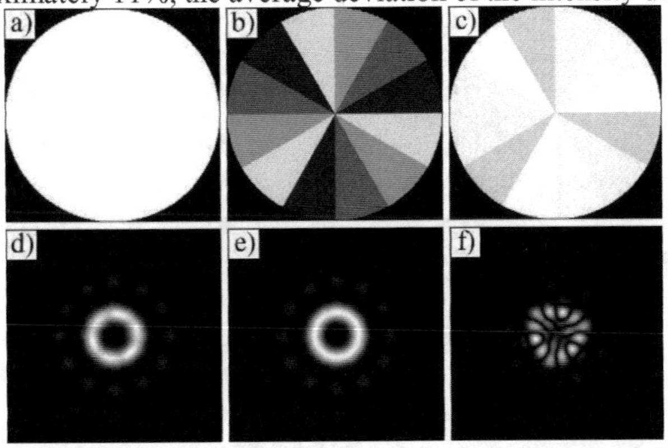

Fig. 1. a) initial distribution of the light field amplitude, b) phase delay profile, c) amplitude of the light field obtained by the polarizer, d) intensity distribution in the focal plane, e) required intensity distribution (without amplitude modulation), f) deviation of the obtained intensity distribution from the desired normalized by its maximum value.

Thus it can be concluded that the resulting amplitude modulation has no significant effect on the formation of light beams with nonzero angular momentum in the form of specified curves. The results of the study indicate the possibility of constructing phase SLM-based FLC with modulation frequencies above 0,5-1 kHz.

The research is carried out under the financial support of RFBR (projects No. 16-42-630773 and 16-29-14012 ofi-m).

References

[1] E.P Pozhidaev et al., Phys. Rev. E **87**, 052502 (2013)

[2] S.P. Kotova et al., Phys. Rev. E **92**, 062502 (2015)

EPJ Web of Conferences **161**, 01008 (2017) DOI: 10.1051/epjconf/201716101008
PECS-2017

Generation of broadband single-photon states in photonic-crystal fibers with flattened dispersion profiles

Ilya Chuprina[2], *and Ilnur* Latypov[1,2,*]

[1]Kazan Quantum Center Kazan National Research Technical University after A.N. Tupolev-KAI, 18a Chetaeva str., Kazan, 420126 Russia
[2]Kazan E. K. Zavoisky Physical-Technical Institute, 10/7 Sibirsky tract Str., Kazan, 420029 Russia

Abstract. We theoretically analyze the capability of generating broadband single-photon states in photonic crystal fibers (PCF) formed by spontaneous four-wave mixing (SFWM) using femtosecond laser pulses. The design of PCF structures with the flat dispersion profile is proposed. The characteristics of the correlated photon pairs with a broadband spectrum and low spectral correlation for these structures are calculated.

1 Introduction

Spontaneous four-wave mixing (SFWM) and spontaneous parametric down-conversion (SPDC) is a powerful tool for generating a wide class of non-classical states, in particular the single-photon or two-photon states [1, 2]. Such states are the basis for long-distance quantum communication, quantum memory and quantum computing protocols. Specific characteristics of such sources determine possibility of its application [3]. The main advantages of these methods include high stability of obtained states in time, the possibility to generate spectrally limited pulses at room temperatures, the possibility to adjust optical wavelength and the spectrum width as well as small influence of multi-photon states. The basic disadvantages of the methods are the random nature of single-photon generation, that leads to a probability of obtaining the two-photon state, and a correlation between the signal and idler photons. An important characteristic of single-photon pulse is the width of the spectrum. It determines the efficiency of interaction with the device quantum information science. An important characteristic of a single-photon pulse is the spectrum line width. It determines the efficiency of interaction with the quantum information device. For example, for cold atomic systems it is required to have a narrowband single-photon state with a spectral line width about 100 MHz. However, for some problems of interest one would need the spectral limited broadband single-photon pulses [4]. Broadband single-photon sources based on spontaneous four-wave mixing with fiber structures are of interest for high-speed transmission of information and high agreement with fiber networks. In this paper we investigate the possibility of generation such states using photonic-crystal fibers. Specific

* Corresponding author: bibidey@mail.ru

design structures of such fibers allow us to obtain a flat profile of dispersion at wavelengths of pump, signal and idler photons.

2 Results

In order to determine spectrum and correlation characteristics for the SFWM photon fields it's necessary to determine the exact values of the refraction index and dispersion of PCFs at its length, that take part in the parametric processes. Usually it is not possible to get analytical solution for the complex internal structure of PCFs to determine its characteristics and to find fundamental modes. That's why computations of the effective refraction index, dispersion and other characteristics of PCF are provided by means of a numerical solution of Maxwell's equations. In the study, simulations were carried out in Lumerical software.

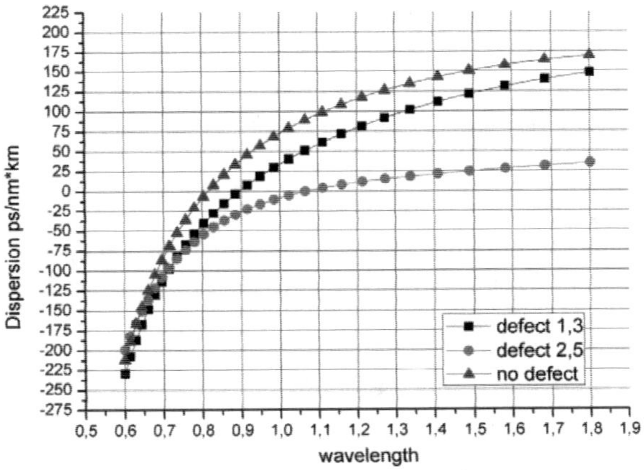

Fig. 1. . The dependence of the fiber dispersion for different structures on the wavelength. Fiber with "no defect" have zero dispersion wavelength at 800 nm. Fiber with "defect 1,3" and "defect 2,5" have circle air hole in core with diameter 260 and 500 nm.

We proposed a new design of photonic crystal fiber structure with the flattened dispersion profile optimized for generation of broadband single photon states in SFWM process using ultrashort laser pulses. Calculation of synchronism conditions showed a significant broadening of the spectrum compared to the standard photonic crystal fiber with low spectral correlation between the signal and idler photon.

The reported study was funded by RFBR according to the research projects № 16-02-00972.

References

[1] S.A. Castelletto, R.E. Scholten, Eur. Phys. J. **41(3)**, 181 (2008)

[2] C.J. Chunnilall et al., Opt. Engineering **53(8)**, 081910 (2014)

[3] A. Migdall, S.V. Polyakov, J. Fan, J.C. Bienfang, *Single-Photon Generation and detection* (Elsevier, Oxford, 2013)

[4] M.R. Sprague, P.S. Michelberger, T.F.M. Champion, D.G. England, J. Nunn, X.-M. Jin, W.S. Kolthammer, A. Abdolvand, P.St.J. Russell, I.A. Walmsley, Nature Photon. **8**, 287 (2014)

Coherent phonons in Bi_2Se_3 induced by a powerful THz pulse

Alexey Melnikov[1,*], *Kirill* Boldyrev[1] , *Yuri* Selivanov[2], *Victor* Martovitskii[2], *Sergey* Chekalin[1], and *Evgeni* Ryabov[1]

[1]Institute of Spectroscopy RAS, 108840 Troitsk, Moscow, Russia
[2]P.N. Lebedev Physics Institute RAS, 119991 Moscow, Russia

Abstract. Using the electro-optic detection method we have detected coherent phonons of E_g^1, E_u^1, E_g^2, and A_{1g}^2 symmetry generated in a crystalline film of Bi_2Se_3 by a powerful single-cycle THz pulse. Coherent excitation of Raman active modes is interpreted as a result of three- and four-phonon interactions in the anharmonic crystal lattice of Bi_2Se_3.

In the present work we studied the ultrafast structural response of a crystalline Bi_2Se_3 film to intense picosecond terahertz pulses. The sample of Bi_2Se_3 under study was an MBE-grown single-crystal film of high structural quality with a thickness of 24 nm on a (111)-oriented BaF_2 substrate covered by a 28 nm thick protective layer of BaF_2. The sample was excited by THz pulses with duration of about 1 ps and peak electric field strength reaching ~ 1 MV/cm. Spectral width of the pulses was ~ 2 THz and the central frequency ~ 1.5 THz. Thus, the IR active phonon mode E_u^1 of Bi_2Se_3 at 67 cm^{-1} (2 THz) was resonantly excited. We detected anisotropic changes of the refractive index of the Bi_2Se_3 film induced by the electric field of the THz pulse. To do that we used a probe laser pulse at 800 nm with a duration of 30 fs. The intensities of its orthogonal polarization components were measured after the pulse propagated through the sample. We registered the dependence of the corresponding ratio I_x/I_y on the delay time between pump and probe pulses. In Fig. 1(a) we show a typical ultrafast response (decay trace) of the Bi_2Se_3 film to the THz pulse. As follows from the FFT spectra of three decay traces measured in succession (Fig. 1(b)) the oscillating signal contains four frequencies of 1.1, 2.05, 4.05, and 5.3 THz (maximal error ~ 0.05 THz) that correspond to optical phonon modes of E_g^1, E_u^1, E_g^2, and A_{1g}^2 symmetry. These oscillations are caused by coherent vibrations of Bi_2Se_3 atoms (coherent phonons), that are induced by the pump THz pulse. The observed generation of coherent phonons cannot be explained using ISRS [1] or DECP [2] theories because the duration of the pump THz pulse is comparable to or larger than the periods of optical phonon modes in Bi_2Se_3. The frequencies of E_g^2 and A_{1g}^2 modes are situated outside the THz pulse spectrum and its direct action on them is excluded. Thus, to explain this unusual effect one should take into accounts higher order processes.

* Corresponding author: melnikov@isan.troitsk.ru

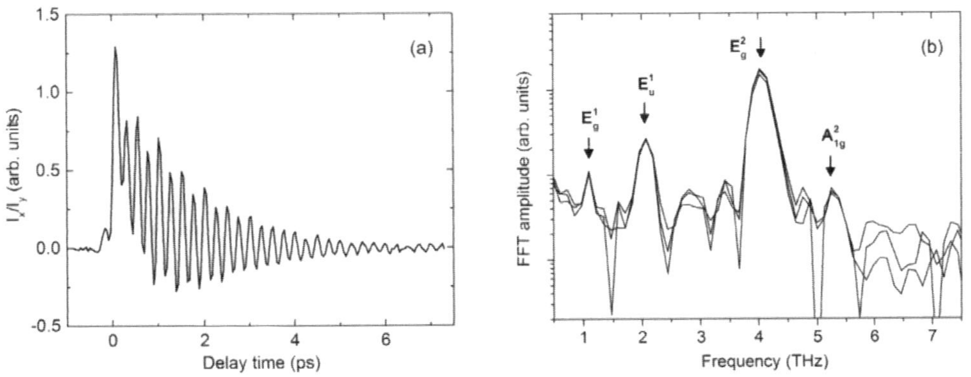

Fig. 1. (a) - Ultrafast response of Bi_2Se_3 to the THz pulse. (b) - FFT spectra of three decay traces measured in succession (vertical axis is in the logarithmic scale).

Recently similar experiments were described in [3] and [4], where the generation of E_g^2 mode was interpreted as a result of vibrationally induced symmetry breaking in the bulk and sum frequency Raman scattering respectively. In contrast to those works we argue here that the coherent excitation of Raman active phonons in Bi_2Se_3 by the THz pulse is the result of crystal lattice anharmonicity and multiphonon processes somewhat similar to "ionic" Raman scattering proposed in [5]. Thus, the generation of E_g^2 phonons is explained as resonant excitation of the Raman active mode by the second harmonic of nonlinear E_u^1 oscillator, the fundamental frequency of which is approximately half that of the E_g^2 mode. In this case the nonlinearity is the consequence of the large amplitude of E_u^1 vibrations and the cubic lattice anharmonicity (three-phonon processes with two E_u^1 phonons and one E_g^2 phonon). At the same time the generation of E_g^1 and A_{1g}^2 phonons can be caused by quartic lattice anharmonicity that provides the frequency combination $2\Omega(E_u^1)=\Omega(A_{1g}^2)-\Omega(E_g^1)$. The proposed interpretation was further supported by experiments with nonresonant radiation and by experiments, in which the dependence of the signal on THz field strength and sample orientation was measured.

This work was supported by the Ministry of Education and Science of the Russian Federation (project # RFMEFI61316X0054).

References

[1] L. Dhar, J. A. Rogers, K. A. Nelson, Chem. Rev. **94**, 157 (1994)

[2] H.J. Zeiger, J. Vidal, T.K. Cheng, E.P. Ippen, G. Dresselhaus, M.S. Dresselhaus, Phys. Rev. B **45**, 768 (1992)

[3] P. Bowlan, J. Bowlan, S.A. Trugman, J.-X. Zhu, N.J. Hur, A.J. Taylor, D.A. Yarotski, and R.P. Prasankumar, Optica **4**, 383 (2017)

[4] L. Braun, *Electron and phonon dynamics in topological insulators at THz frequencies* (Freie Universität Berlin, Germany, 2016)

[5] M. Först, C. Manzoni, S. Kaiser, Y. Tomioka, Y. Tokura, R. Merlin, and A. Cavalleri, Nature Phys. **7**, 854 (2011)

Magneto-optical properties of the photon echo: status of research and prospects of application

I.I. Popov[1,*], A.U. Bahadurov[1], N.S. Vashurin[1], E.A. Vinogradov[2], V.O. Kompanets[2], S.V. Chekalin[2], and S.E. Putilin[3]

[1]Volga State Technological University, 3 Lenin sq., Yoshkar-Ola, 424001 Russia
[2]Institute of Spectroscopy of the Russian Academy of Sciences, 5 Fizicheskaya str, Moscow, Troitsk, 108840 Russia
[3]St. Petersburg National Research University of Information Technologies, Mechanics and Optics, 49 Kronverksky Prospect, St. Petersburg, 197101 Russia

Abstract. The results of the study of the polarization properties of the stimulated photon echo (SPE) excited at room temperature on free excitons and exciton states localized on surface defects of textured thin films are reported. The results of observation and investigation of the effect of the non-Faraday rotation of a femtosecond SPE in a textured thin film are presented. The physical processes occurring under the Faraday effect and the effect of the non-Faraday rotation of the photon echo (PE) polarization plane are compared.

When the PE is formed in the presence of a longitudinal homogeneous magnetic field, the effect of the non-Faraday rotation of the PE polarization plane takes place. This effect differs fundamentally from the Faraday effect in the physics of the occurring processes and the magnitude of the rotation angle, which is three times bigger in exponent part than the Faraday rotation at the same values of the intensity of the longitudinally applied uniform magnetic field. The angle of the non-Faraday rotation of the PE polarization plane depends on the g factor of the quantum transition of the resonant medium, on the magnitude of the applied magnetic field H and on the time intervals separating the exciting pulses. The PE is formed during the optical transient process when the exciting laser radiation is applied to a resonant medium. The difference between these effects is as follows.

When the quantum system is excited by pulses with a duration much shorter than the relaxation times of the resonant medium equally polarized pseudoelectric dipoles are produced on all particles. The process of dephasing (phasing) of these dipoles occurs at different phase velocities for the upper (dextrorotatory) and lower (left-handed) components of the hyperfine structure of the excited level, which is split due to the Zeeman effect. The upper component increases, while the lower component decreases the phase velocity of the precession of the dipole. The resulting phase velocity difference, which leads to the rotation of the polarization plane of the echo signal, increases with increasing magnetic field strength and the time of dephasing (phasing) of the dipoles, given by the value of the time intervals

* Corresponding author: Popov@volgatech.net

separating the exciting pulses. The non-Faraday rotation occurs both for the PE and for the SPE.

For the first time, the non-Faraday rotation of the SPE at room temperature in the nanosecond range of durations is performed in molecular iodine vapor [1]. Since we first obtained femtosecond signals of the primary photon echo PE and SPE on free excitons and excitons localized on surface defects assembled into nanoscale clusters, we had the opportunity to conduct a search for the non-Faraday rotation of the PE polarization plane in an exciton gas. Because of the high temperature, the thermal energy of the excitons is comparable to their binding energy, so a significant part of the excitons localized in the cluster recombine, and the concentration of the remaining excitons decreases to a concentration typical for the gas medium. This circumstance makes it possible to effectively use the available theory of the effect of the non-Faraday rotation of the polarization plane of the PE in a gas very effectively for processing the experimental results obtained. The difference in the defects entering the cluster ensures an inhomogeneous broadening of the exciton gas, which is necessary for the formation of a photon echo. The photon echo is formed during the transient process of formation of the exciton state, that is, on superposition states of pseudoelectric dipoles created by an electron having a Coulomb binding with a semiconductor hole.

A series of our experiments with femtosecond pulses on the study of the polarization properties of SPE in the presence of a longitudinal homogeneous magnetic field includes:

1. Detection of the effect of the non-Faraday rotation of the polarization plane of the SPE.
2. The results of the research of this effect allow to register the femtosecond time intervals at room temperature on-line.
3. Detection of the inhomogeneous distribution of localized exciton states by measuring the g factor of quantum transitions from the repetition period of the modulation on the slope of the SPE curve as a function of the time interval between the exciting pulses of τ_{23} for several discrete values of the time interval τ_{12}. These results can be used as the basis for developing a new method of femtosecond magneto-optical spectroscopy based on SPE.
4. Detection of a long-lived femtosecond PE on localized exciton states at room temperature.

The principal difference between the last experiment on long-lived memory in the works of I.A. Akimov and his coauthors [2,3] is the following (apart from the fact that the colleagues observed a nanosecond PE at helium temperature, and not a femtosecond PE, like us, and at room temperature). In their works, the Larmor precession under the action of a transverse magnetic field was used to switch over the back of the object on which the PE was formed. In our research, the Larmor precession, under the action of a longitudinal magnetic field, influenced the phase velocity of the precession of the pseudoelectric dipole with a different contribution for the components of the hyperfine structure of the excited quantum level, from which the degeneracy was removed due to Zeeman splitting.

References

[1] I.S. Bikbov, I.I. Popov, V.V. Samartsev, I.V. Yevseyev, Laser Phys. **5**(3), 580 (1995)

[2] L. Langer, S.V. Poltavtsev, I.A. Yugova et al., Phys. Rev. Lett. **109**, 157403 (2012)

[3] L. Langer, S.V. Poltavtsev, I.A. Yugova et al., Nature Photon. **8**, 851 (2014)

EPJ Web of Conferences **161**, 01011 (2017) DOI: 10.1051/epjconf/201716101011
PECS-2017

Spectral manifestation of distorted forms metalloporphyrins at low temperatures

Aleksander Starukhin[1,*], *Aleksander* Gorski[2], and *Michal* Kijak[2]

[1]B.I. Stepanov Institute of Physics, National Academy of Sciences of Belarus, Minsk, Belarus
[2]Institute of Physical Chemistry, Polish Academy of Sciences, Warsaw, Poland

Abstract. Different spectral forms of a set of metalloporphyrins with Mg (II), Zn (II), Pd (II) and Pt (II) ions in solid solutions (Shpol'skii matrices, solid tetrahydrofuran and solid rare gas matrices) have been detected in the absorption and luminescence spectra at cryogenic temperatures. The spectral manifestation of a planar form, as well as two distorted forms for Mg- and Zn-porpyrins was formed in the ground electronic state at low temperatures. An existence of two spectral forms of Pd and Pt-porphyrins was observed in their phosphorescence spectra. In the case of Mg- porphyrin the spectral shift between positions of 0-0 transitions the planar and first distorted spectral forms is about 220 cm^{-1}.

Metallocompexes of porphyrins are biologically relevant compounds known for many years. It is well known, that metallocomplexes of porphyrins demonstrate planar structures of high symmetry at room temperature in neutral solvents. However, under certain conditions, the planar structure of the porphyrin macrocycle may be distorted, that leads to corresponding spectral manifestations of the nonplanarity effects. First, the method of resonance Raman spectroscopy was used for recording and analysing of distorted structures of metalloporphyrins [1]. Nevertheless, the mentioned method does not have selectivity for analysing of the manifestation of several types of distorted forms together. In contrast, the methods of luminescent spectroscopy may be effectively used in a cases of stabilization of various (including liganded) spectral forms at low temperatures. Fluorescence line narrowing procedure at low temperatures [3] permits excites selectively and registers the highly resolved spectra of individual conformers and assigns the frequencies of normal modes.

In this report we were discussing progress in study of metallocomplexes of porphine (P), octamethylporphyrin and octaethylporphyrin in different solid matrices at cryogenic temperatures. Spectral manifestation of different forms have been investigated and characterized by methods of luminescent and fine line spectroscopy.

For example, three spectral forms with positions of 0-0 transition located: at 552,4 nm (form F0), 562,4 (form FI) and 566,5 (form FII) have been detected in fluorescence spectra of Zn-porphyrin (Zn-P) in solid Ar at 4.2 K. A spectral manifestation of above mentioned forms is due to are stabilized of planar (form F0) and two different distorted FI and FII conformations of Zn-P in low temperatures matrices. Forming of these forms in the ground state has been proved early for Mg-P and Zn-P upon selective detection of fluorescence

* Corresponding author: a.starukhin@ifanbel.bas-net.by

excitation and fluorescence spectra (3). For Zn-P distorted forms have been detected in the fluorescence and phosphorescence spectra simultaneously. The manifestation of the two forms spectra of the Pd- and Pt-P has been observed in the phosphorescence and excitation of fluorescence and phosphorescence spectra [4] with spectral shift of the 0-0 positions about 78 cm^{-1} and 38 cm^{-1}, respectively.

One of the most fascinating results is the detection of absorption (fluorescence excitation) fine line spectra of Mg-P and Zn-P of Soret band of (Fig. 1).

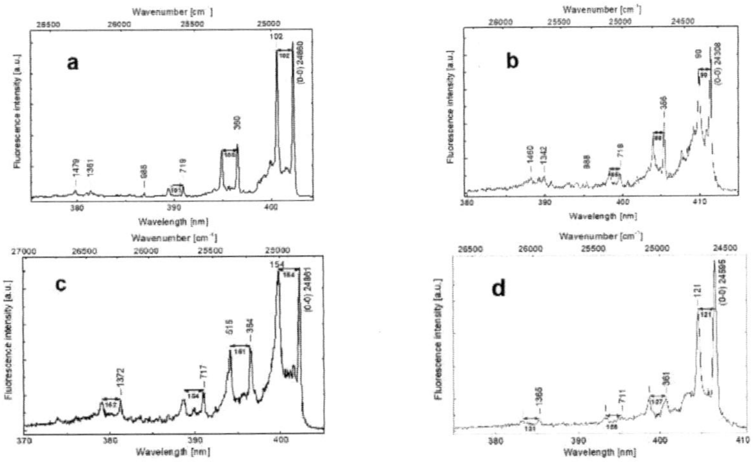

Fig. 1. Fluorescence excitation spectra of Mg-P and Zn-P in the range of Soret band of Form I (**a** - Mg-P, **c** - Zn-P) and Form II (**b** - Mg-P, **d** - Zn-P in solid matrix of tetrahydrofurane (THF) at 4,2 K

The set of vibrational modes of the excited state (Soret band) was detected, e.g. for FI form of Mg-P vibronic lines with frequencies 360, 719, 986, 1361 and 1479 cm^{-1} (Fig. 1a). Interestinglly, that vibronic bands are split into doublets with a constant separation, specific for a given form and type of a central ion. Density Functional Theory (DFT) and time dependent DFT methods were used for moddeling of the ground and electronically excited states of investigated mettalocomplexes. Virbational analysis was performed to explaine observed spectral data. Based on that, an assignment of experimentally observed forms was proposed. The FI form was attributed to a complex of metalloporphyrin with one extra ligand placed above its plane. The FII form corresponds to a structure with two additional ligands located on opossite sides of the porphyrin macrocycle. Jahn-Teller distortions of studied molecules in Q and Soret states and resulting band splitting has been also analysed.

This work was supported by the BRFFI Grant No. F16R-084 and Horizon 2020 MSCA RISE Grant 645628 .

References

[1] J.A. Shelnutt et al., Chem. Soc. Rev. **27,** 31 (1998)

[2] R.I. Personov, E.I. Al'shits, L.A. Bykovskaya, Opt. Commun. **7**, 169 (1973)

[3] A.S. Starukhin, A.M. Shul'ga, Opt. and Spectrosc. **98**, 780 (2005)

[4] A. Starukhin, M. Kruk, R. Czerwieniec, J. Lumin. **128,** 531 (2008)

DC Stark addressing for quantum memory in Tm:YAG

Konstantin Gerasimov[1], *Mansur* Minnegaliev[1], *Ravil* Urmancheev[1,*], and *Sergey* Moiseev[1]

[1]Kazan Quantum Center, Kazan National Research Technical University named after A.N. Tupolev – KAI, Kazan, 420111 Russia

Abstract. We observed a linear DC Stark effect for $^3H_6$ – $^3H_4$ optical transition of Tm^{3+} ions in $Y_3Al_5O_{12}$. We observed that application of electric field pulse suppresses the two-pulse photon echo signal. If we then apply a second electric pulse of opposite polarity the echo signal is restored again, which indicates the linear nature of the observed effect. The effect is present despite the D_2 symmetry of the Tm^{3+} sites that prohibits a linear Stark effect. Experimental data analysis shows that the observed electric field influence can be attributed to defects that break the local crystal field symmetry near Tm^{3+} ions. Using this effect we demonstrate selective retrieval of light pulses in two-pulse photon echo.

The development of optical quantum technologies such as quantum computer and long-distance quantum communications stimulates the creation of quantum memories, devices that would broaden their possible applications. One of the promising media in the field of quantum memories is the multi-atomic ensembles that allow one to strengthen the interaction with weak photon fields. Rare-earth ions stand out among such systems due to remarkable combination of long T_2 coherence times and broad inhomogeneous optical transitions and possibility to realize a long-time storage on electron-nuclear spin sub-levels.

Tm^{3+}:YAG crystal has been extensively investigated due to its laser applications and its 793 nm wavelength transition $^3H_6$ – $^3H_4$ that lies in the third optical fibre transparency window. Tm ions in this material substitute Y ions in crystallographic sites with D_2 local symmetry. Linear Stark effect is absent in central symmetries [1] such as D_2. However in this work we demonstrate a quasi-linear Stark effect during the two-pulse photon echo formation at 4K temperature.

In the experiments wave vectors of the incoming pulses were oriented along the $[1\bar{1}0]$ crystallographic axis. Both pulses were polarised along the [001] axis. In these conditions all exited thulium atoms had the same Rabi frequency. We observed the suppression of the photon echo after the applied electric field pulse. The echo was suppressed completely after the pulse with ≥ 5 μs duration and $E \geq 500$ V/cm field amplitude. Keeping in mind that D_2 symmetry prohibits linear Stark effect we first attributed this to quadratic Stark effect. But actually when we applied a second pulse of the same duration and amplitude but opposite polarity the echo was revived with roughly the same amplitude. The revival also took place if the second Stark pulse had the same sign but was sent between the π-pulse and the echo.

* Corresponding author: ravil@kazanqc.org

During the experiments we saw that the effect is present in 0 magnetic field as well as in 60 mT field oriented along the [001] crystallographic axis. These two revival conditions indicate that the observed Stark effect is linear despite the fact that it shouldn't be observed in the D_2 site symmetry. This combination of suppression and revival of the echo may allow one to realize an addressing photon echo quantum memory protocol [2].

We attribute the linear Stark effect to the disturbances of the crystal field around the Tm ion sites that break the D_2 symmetry. That gives each ion a phase shift different from the other ions. We also saw that the effect's magnitude is proportional to the frequency shift between the laser pulses carrier frequency and the inhomogeneous line centre.

Using the observed effect we realized address readout from the optical memory cell. We retrieved desired light pulses using electric pulses with amplitude of 500 V/cm and 5 us duration. Fig. 1 shows the effect of different combinations of electric pulses and their compensating counterparts on the optical memory writing and reading process. Using the corresponding electric pulses we can suppress any one of the pulses. Using the additional π-pulse we can also retrieve pulses in the reversed order.

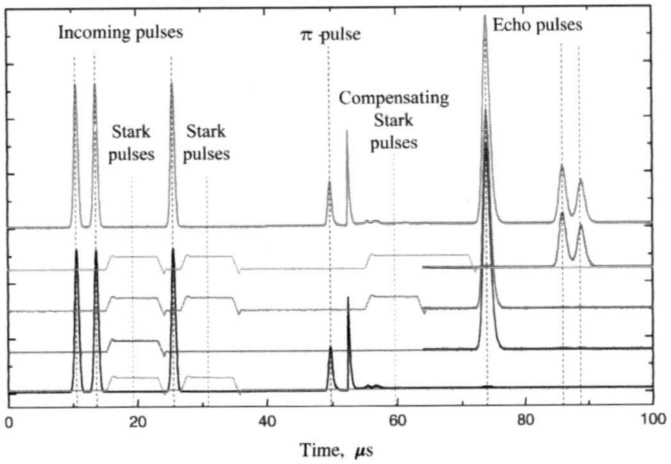

Fig. 1. Controlled retrieval of the photon echo signals using Stark pulses in Tm^{3+}:YAG crystal. The darker solid curves show the incoming pulse sequences and echo pulses. The lighter solid lines show Stark pulses that suppress the echo and compensating pulses that revive desired echo pulses. There are always three incoming pulses – one pair and one single pulse. After them at 50 us time comes the rephasing π-pulse. Its tail can be seen at roughly 53 us at the lowest and the highest curves. The highest curve show the retrieval of all pulses and lowest curve – their full suppression. The second from the bottom curve shows the retrieval of just the single pulse. Above that – the suppression of all pulses and then revival of the single pulse using the additional compensating Stark pulse. Above that is the revival of both single pulse and the pair of pulses.

This research has been supported by the Government of Russian Federation, project no. 14.Z50.31.0040, Feb. 17, 2017.

References

[1] R.M. Macfarlane, J. Lumin. **125**, 156 (2007)

[2] K.I. Gerasimov, M.M. Minnegaliev, S.A. Moiseev, R.V. Urmancheev, T. Chanelière and A. Louchet-Chauvet, Opt. Spectrosc. **123**(2), 211 (2017)

Integrable models of quantum optics

Vladimir Yudson[1,2,*] and *Aleksander* Makarov[2,3]

[1]National Research Institute Higher School of Economy, 20 Myasnitskaya Str., Moscow, 101000 Russia
[2]Institute of Spectroscopy RAS, 5 Fizicheskaya Str., Troitsk, Moscow 142190 Russia
[3]Moscow Institute of Physics and Technology, 9 Institutskiy Pereulok, Dolgoprudny, 141700 Russia

Abstract. We give an overview of exactly solvable many-body models of quantum optics. Among them is a system of two-level atoms which interact with photons propagating in a one-dimensional (1D) chiral waveguide; exact eigenstates of this system can be explicitly constructed. This approach is used also for a system of closely located atoms in the usual (non-chiral) waveguide or in 3D space. Moreover, it is shown that for an arbitrary atomic system with a cascade spontaneous radiative decay, the fluorescence spectrum can be described by an exact analytic expression which accounts for interference of emitted photons. Open questions related with broken integrability are discussed.

The simplest model of quantum optics - a two-level atom interacting with a single mode quantum electromagnetic field (Janes-Cummings model [1]) allows one to find exact states of the system and to describe its dynamics. For a multimode field the problem becomes nontrivial. Examples of a multimode field are a field propagating in a long waveguide, a field in an open 3D space, etc. Solution of the Schrödinger equation for a multi-atom - multi-photon system is not possible in general. However, this is possible for some particular models of quantum optics which we describe below.

1. A system of two-level atoms in a 1D waveguide with chiral photons described by the Hamiltonian (a quantum generalization of the classical model)

$$H = -i\int dx\, a^{+}(x)\partial_{x}a(x) - \sqrt{\gamma}\sum_{j}\left[s_{j}^{+}a(x_{j}) + s_{j}^{-}a^{+}(x_{j})\right] \quad , \qquad (1)$$

where $a(x)$ is the photon annihilation operator at point x and spin-1/2 operator $\vec{s}_{j}$ describes a two-level atom located at the point x_{j}. The exact eigenstates of this Hamiltonian for an arbitrary number of photons and excited atoms were constructed in [3] with the use of Bethe Ansatz [4]. The chirality (unidirectional propagation) of photons is the key condition for the integrability of this model.

2. A system of M identical two-level atoms located in a spatial region (of a size small as compared with the resonant wavelength) and interacting with electromagnetic field in 3D space (Dicke model [5]). As shown in [6], this 3D model can be mapped on an effective 1D

* Corresponding author: yudson@isan.troitsk.ru

model Eq. (1), but with chiral photons interacting with a spin M/2 located at the point $x = 0$. Such a model also allows one to construct exact multi-photon eigenstates [6]. Knowledge of the full set of eigenstates and an efficient method developed in [7] for summation over so called string configurations (correlated packets of several photons) allows one to describe dynamics of an arbitrary initial state of the many-body atom-photon system. In particular, it has become possible to describe not only the cascade decay dynamics of the atomic system down to the ladder of Dicke's states, but also to describe the time evolution of multiphoton functions and to calculate the superfluorescence spectrum.

3. Exactly, albeit in a more cumbersome way, one can solve also a problem of N photons interacting with an ensemble of M closely located two-level atoms in the usual (non-chiral) 1D waveguide [8]. This is possible because there are no photon returns to the atomic system in this model. Such returns are present already in a simplest system of separated atoms in the non-chiral waveguide; they essentially influence on the system dynamics and, for instance, lead to existence of dark states [9].

4. An exact description is possible also for an arbitrary atomic system with a cascade system of optical transitions. The absence of branching and returns allows one to find the exact solution of the Schrödinger equation for the amplitudes of the states with arbitrary numbers of emitted photons. In this way one can obtain a general expression for the spectrum of the cascade emission [10]:

$$S(\omega) = \frac{1}{2\pi i}\sum_{l=1}^{N}\frac{1}{\Gamma_{l,l-1}}\sum_{k=1}^{l}\prod_{j=k}^{l}\frac{\Gamma_{j,j-1}}{\omega-(E_j-E_{j-1})-\frac{i}{2}(\Gamma_{j,j-1}+\Gamma_{j-1,j-2})} + c.c. \qquad (2)$$

where E_j is the energy of the state $|j\rangle$, $\Gamma_{j,j-1}$ - the rate of the transition $|j\rangle \to |j-1\rangle$. This expression accounts also for the case of close frequencies of some transitions (so that the emitted photon can interfere). For a particular case of the cascade emission in the Dicke model, Eq. (2) coincides with that calculated in [6] with the use of Bethe Ansatz; the transition rates are $\Gamma_{j,j-1} = j(N-j+1)\Gamma$.

References

[1] E.T. Jaynes, F.W. Cummings, Proc. IEEE **51**, 89 (1963)
[2] J.C. MacGillivray and M.S. Feld, Phys. Rev. A **14**, 1169 (1976)
[3] V.I. Rupasov, V.I. Yudson, Sov. Phys. JETP **59**(3), 478 (1984)
[4] H. Bethe, Z. Phys. **71**, 205 (1931)
[5] R.H. Dicke, Phys. Rev. **93**, 99 (1954)
[6] V.I. Rupasov, V.I. Yudson, Sov. Phys. JETP **60**, 927 (1984)
[7] V.I. Yudson, Sov. Phys. JETP **61**, 1043 (1985)
[8] V.I. Yudson, P. Reineker, Phys.Rev. A **78**, 052713 (2008)
[9] A.A. Makarov, V.S. Letokhov, JETP **97**, 688 (2003)
[10] A.A. Makarov, V.I. Yudson, Phys. Rev. A **89**, 053806 (2014)

EPJ Web of Conferences **161**, 02001 (2017) DOI: 10.1051/epjconf/201716102001
PECS-2017

Higgs field and cosmological parameters in the fractal quantum system

Valeriy Abramov*

Donetsk Institute for Physics and Engineering named after A.A. Galkin, Ukraine

Abstract. For the fractal model of the Universe the relations of cosmological parameters and the Higgs field are established. Estimates of the critical density, the expansion and speed-up parameters of the Universe (the Hubble constant and the cosmological redshift); temperature and anisotropy of the cosmic microwave background radiation were performed.

The hypothesis of the hierarchical structure of the Universe (HHSU) became the basis for fractal cosmology [1]. The development of the technique of spatial scanning made it possible to determine the speed of expansion of the Universe, the cosmic distances to the far Galaxies with great accuracy [2]. As separate elements of large-scale fractal structures of the Universe (SELFSU), Galaxies, clusters and superclusters of Galaxies, the largest supercluster – Great Attractor, walls, filaments, voids are considered [1]. HHSU makes it possible to use models of active nanoelements (fractal dislocations, quantum dots with variable parameters [3]) to describe SELFSU [4]. The aim of the work is to establish the connections between cosmological parameters and the Higgs field within the framework of the fractal model of the Universe. For the unit cell (such as the effective atom and the Higgs boson) model relationships, that relate the Hubble constant, the cosmological redshift, the temperature and the anisotropy of the relic radiation are obtained. The shift of the cosmic microwave background (CMB) frequency is described by equations

$$\delta\nu_\gamma=4Q_{H2}\,\gamma\,u_{\Omega1}^2/N_0' ;\ \delta\nu_\gamma'=N_0'\,\delta\nu_\gamma/N_{02}' ;\ u_{\Omega1}^2=(\gamma-|\xi_q|)/2\gamma ;\ \xi_q=\nu_{r1}-\nu_{r2} ;\ \Omega_{c1}'=4|\lambda|^2 N/|\xi_q|^2 . \quad (1)$$

Using from [4] the values $Q_{H2}=1/3$, $\nu_{r1}=160.3988698\,\text{GHz}$ (the maximum of the CMB frequency), N_0', N_{02}', $\nu_{r2}=142.8161605\,\text{GHz}$ (the shifted radiation frequency), we find the frequency shifts $\delta\nu_\gamma=3.3920592\,\text{kHz}$, $\delta\nu_\gamma'=3.4281246\text{kHz}$ and the density of cold dark matter $\Omega_{c1}'=0.228071512$. The value $\delta\nu_\gamma'$ (received with the Higgs field) agrees with the experiment on shifting the NMR frequency (3.42 kHz) for superfluid helium-3 [5]. Nonlinear equations (determine the parameters of the CMB) for functions Q_G, B_G are

$$Q_G=Q_H(1-Q_H)^2/(1+Q_H(1-Q_H)^2) ;\quad B_G=1-2Q_G ;\quad V_Q=dQ_G/dQ_H ;\quad A_Q=dV_Q/dQ_H . \quad (2)$$

* Corresponding author: vsabramov@mail.ru

 DOI: 10.1051/epjconf/201716102001

The dependences of these functions on Q_H are given in Fig. 1. Variant $V_Q > 0$ and $A_Q > 0$ corresponds to the accelerated expansion of the Universe [4]. In model I [4] $Q_{H0} = H_{01} / H_{02}$, where H_{01}, H_{02} are the Hubble constants.

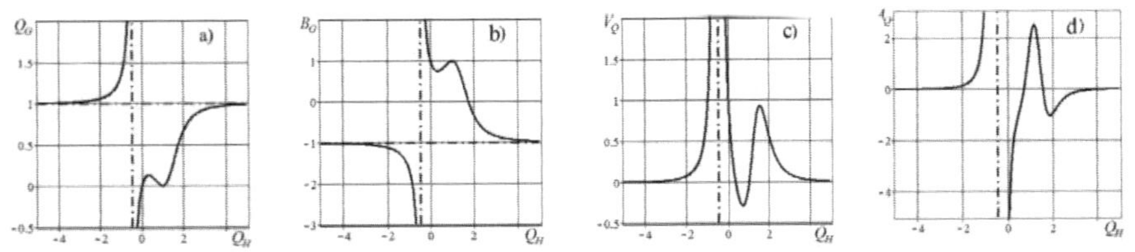

Fig. 1. The behavior of functions Q_G (a), B_G (b), V_Q (c), A_Q (d) on Q_H.

The modeling results of the deformation field (u_2) for a coupled system (dislocation-quantum dot) are given on fig. 2. The function $u_2(n,m)$ is complex with a pronounced stochastic behavior and essentially depends on the relationships between the characteristic dimensions $N_1 \times N_2$ of the investigated region, the values of the semi-axes n_{c2}, m_{c2} at the quantum dot (fig. 2). Fractal holes with proportional semi-axes (fig. 2 a), elongated along the axis On (fig. 2 b) and along the axis Om (fig. 2 c) in elliptical structures are observed inside the region of wave behavior (fig. 2 a,b,c). In this case, a pronounced phenomenon of the anisotropy of the fractal hole appears. The character of the anisotropy for effective damping (fig. 2 d,e,f) agrees with the anisotropy of the fractal hole.

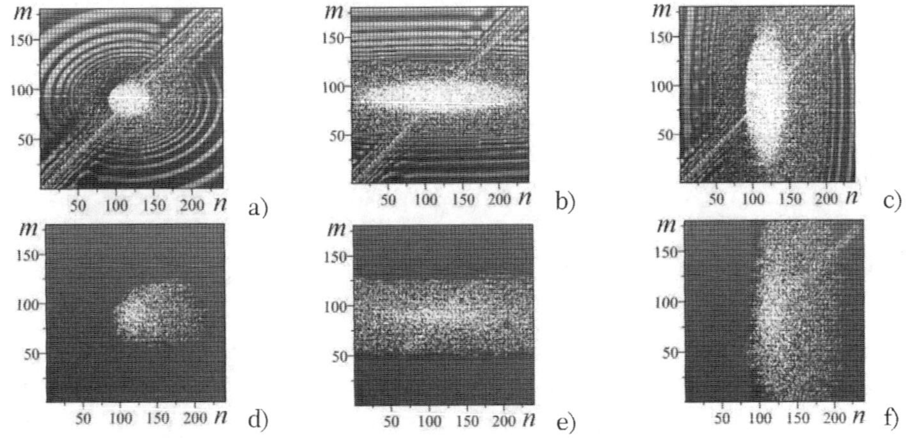

Fig. 2. Cross-sections $\mathrm{Re}(u_2) \in [-10;10]$ (a,b,c), $\mathrm{Im}(u_2) \in [-10;10]$ (d,e,f) (top view) for coupled system with variable modules k_1 for u_2 and semi-axes for quantum dot: (a, d) $3n_{c2}$; $3m_{c2}$; (b, e) $12n_{c2}$; $3m_{c2}$; (c, f) $3n_{c2}$; $12m_{c2}$.

References

[1] B. Novosyadlyy, Universe, Space, Time **6**(143), 4 (2016)
[2] A.G. Riess, A.V. Filippenko et al., Astronomical Journal **116**, 1009 (1998)
[3] V.S. Abramov, Bull. RAS: Physics **80**(7), 859 (2016)
[4] V.S. Abramov, Bull. of Donetsk National University **A**(1), 36 (2017)
[5] *Superfluidity in Helium-3*: *Collection of Articles* (Moscow, Mir, 1977)

EPJ Web of Conferences **161**, 02002 (2017) DOI: 10.1051/epjconf/201716102002
PECS-2017

Towards spontaneous parametric down-conversion at low temperatures

Dmitrii Akatiev[1,*], *Kirill* Boldyrev[2], *Nikolai* Kuzmin[2,3], *Ilnur* Latypov[1], *Marina* Popova[2], *Andrey* Shkalikov[1], and *Alexey* Kalachev[1]

[1]Kazan E. K. Zavoisky Physical-Technical Institute, 10/7 Sibirsky trakt Str., Kazan, 420029 Russia
[2]Institute of Spectroscopy RAS, 5 Fizicheskaya Str., Troitsk, Moscow, 108840 Russia
[3]Lomonosov Moscow State University, GSP-1 Leninskie Gory, Moscow, 119991 Russia

Abstract. The possibility of observing spontaneous parametric down-conversion in doped nonlinear crystals at low temperatures, which would be useful for combining heralded single-photon sources and quantum memories, is studied theoretically. The ordinary refractive index of a lithium niobate crystal doped with magnesium oxide $LiNbO_3$:MgO is measured at liquid nitrogen and helium temperatures. On the basis of the experimental data, the coefficients of the Sellmeier equation are determined for the temperatures from 5 to 300 K. In addition, a poling period of the nonlinear crystal has been calculated for observing type-0 spontaneous parametric down-conversion (ooo-synchronism) at the liquid helium temperature under pumping at the wavelength of λ_p = 532 nm and emission of the signal field at the wavelength of λ_s = 794 nm, which corresponds to the resonant absorption line of Tm_{3+} doped ions.

1 Introduction

Spontaneous parametric down-conversion (SPDC) is the most common way of generating non-classical states of the electromagnetic field, in particular single-photon or two-photon states [1, 2]. When such states are used in the systems of long-distance quantum communications or quantum computing, it may be useful to have the possibility to store them in a quantum memory device immediately after their generation. In this case, the optimal solution might be the combination of the SPDC-based photon source and quantum memory in a single system utilizing a doped nonlinear crystal as an information carrier. Since solid-state quantum memories typically work at liquid-helium temperature, it is necessary that single-photon or two-photon sources do the same. However, to our knowledge, no one has experimentally studied SPDC under such conditions. Furthermore, even temperature dependence of refractive indices of nonlinear crystals that could be used for combining single-photon sources and quantum memories has not been investigated at low temperatures. A promising example is the lithium niobate crystal doped with trivalent thulium ions [3]. In the pre-sent work, we study the possibility of observing SPDC at low temperatures in such nonlinear crystals.

[*] Corresponding author: akatevdmitrijj@gmail.com

2 Basic results

For the present theoretical and experimental study, we have chosen lithium niobate crystal doped with magnesium oxide (5%) $LiNbO_3$:MgO produced by "Labfer". The dependence of the ordinary refractive index on the wavelength has been measured at different temperatures (see Table 1 and Figure 1).

Table 1. The measured deviation of ordinary refractive index $\Delta n = n_T - n_{298\,K}$ from its room temperature value for the lithium niobate crystal doped with magnesium oxide $LiNbO_3$:MgO (5%) for some temperatures and wavelengths.

	λ = 0.53775 μm	λ = 0.812 μm	λ = 1.075 μm	λ = 1.55 μm
T = 4.3 K	-0.0025	-0.0012	-0.006	-0.006
T = 77.4 K	-0.0034	-0.0011	-0.006	-0.006
T = 298 K	0	0	0	0

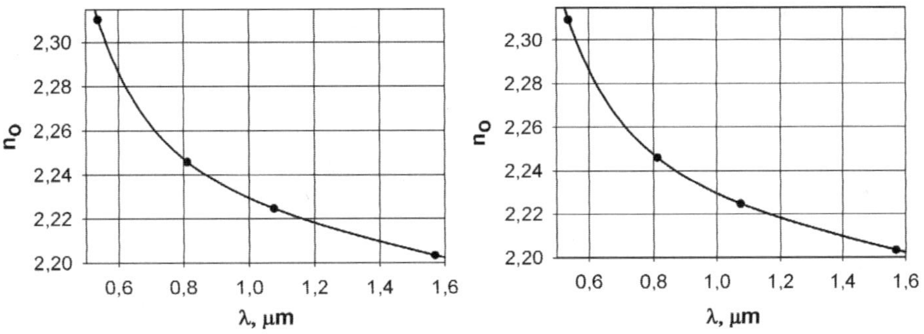

Fig. 1. The dependence of the ordinary refractive index on the wavelength at liquid helium (left) and liquid nitrogen (right) temperature. The circles represent experimental data, while the solid line is the approximation using modified Sellmeier equations.

On the basis of the experimental data, we study the possibility of observing type-0 SPDC at the liquid helium temperature under pumping at the wavelength of λ_p = 532 nm and emission of the signal field at the wavelength of λ_s = 794 nm, which corresponds to the resonant absorption line of Tm^{3+} doped ions. Numerics show that in this case the lithium niobate crystal should have the poling period about Λ = 6.63 μm. Such a regime of SPDC would be useful for combining a heralded single-photon source and quantum memory device in a single crystal, since the wavelength of the signal photons corresponds to the resonant absorption line of Tm^{3+} doped ions. The preliminary experimental work has been carried out [4] which demonstrates the possibility of SPDC in this crystal at temperatures below 0°C.

References

[1] S.A. Castelletto, R.E. Scholten, Eur. Phys. J. **41(3)**, 181 (2008)
[2] C.J. Chunnilall et al., Opt. Engineering **53(8),** 081910 (2014)
[3] C.W. Thiel, Y. Sun, R.M. Macfarlane, T. Böttger, R.L. Cone, J. Phys. B: At. Mol. Opt. Phys. **45**, 124013 (2012)
[4] I.Z. Latypov et al., J. Phys.: Conf. Ser. **859**, 012011 (2017)

 DOI: 10.1051/epjconf/201716102003

Time-frequency mapping of the positional characteristics of single- and multimode wave beam at the lengthy atmospheric path output

Tatiana Arsenyan[2], *Eugeniy* Babanin[2,⋆], *Vitaliy* Kapranov[1], *Rinat* Stryungis[2], *Natalia* Suhareva[2], and *Vjatcheslav* Tugaenko[1]

[1] *S.P. Korolev Rocket and Space Corporation "Energia" (Russian Federation)*
[2] *M.V. Lomonosov Moscow State University (Russian Federation)*

Abstract. The comparative experimental analysis of the single-mode and multimode positional characteristics of the wave beams at the lengthy atmospheric path output was carried out and the results are presented. As the main instrument the method of nonlinear time-frequency mapping was chosen using Wigner function apparatus. The characteristic time-frequency spectra under the conditions of strong, intermediate and weak turbulence of the atmospheric path were defined.

The broad range of challenges suggested by the optical propagation channel demands quick signal correction to compensate the atmosphere induced distortions. In this article the sequential model for the analysis of the positional characteristics of the wave beams using the nonlinear time-frequency mapping is presented.

Let us construct the time-space maps for nonstationary processes of the variations of a wave beam spatial moments on the base of Wigner-Willy representation. This allows circumventing restrictions of the uncertainty principle for time and frequency resolution [1]:

$$WVT_s(t,\omega) = \int_{-\infty}^{\infty} s(t+\tau/2)\cdot s^*(t-\tau/2)e^{-i\omega\tau}d\tau, \quad (1)$$

Here $s(t)$ is the initial signal. The presence of strong interference distortions makes the experimental results more difficult to interpret.

Traditionally for meteomonitoring they use the C_n^2 parameter or the Rytov parameter β^2. There is a three interval classification in radio physical applications that determines the Rytov parameter values corresponding to weak - $[0, 0.3)$, average - $[0.3, 1)$ and strong - $[1, ...)$ turbulence development [2]

The time-frequency spectrum also depends on the atmospheric conditions. The first special moments spectra describe "the translational" beam motion. The spectra of the second moments describe the vibrational modes of the collimated beam and its nondiagonal element $xyVar$ describes "rotatory" movement thus changing the direction of the orientation axis of approximating ellipsoid.

Under the conditions described the sampling rate was equal to 100 Hz. In the case of weak turbulence the main components of the spectrum appear to be in the frequency range up to 5 Hz (Fig.1)). When the Rytov parameter increases up to 2.5 the spectrum occupies all the frequency range observed.

⋆e-mail: babanin.eugen@gmail.com

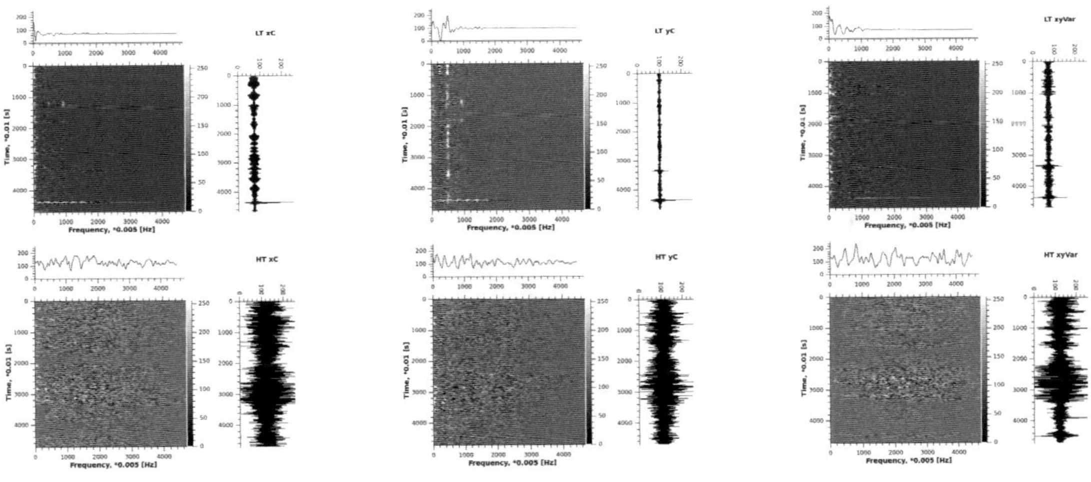

Figure 1. Time-frequency spectra for components of first moment and nondiagonal element of second moment in conditions of low (top three figures) and hight turbulence (bottom three figyres).

A multimode wave beam was used for meteomonitoring tasks and its diameter increased the registration plane size [2]. The time-frequency spectra of the illumination variance presented on Fig.2 were obtained in conditions corresponding to three turbulence levels. The sampling rate was equal to 500 Hz. When the Rytov parameter increases the spectrum occupies the higher frequency range. Moreover for multimode beam the time-frequency spectrum depends on the square frame liner size. When we increase the frame size the spectrum shifts to the low frequency area.

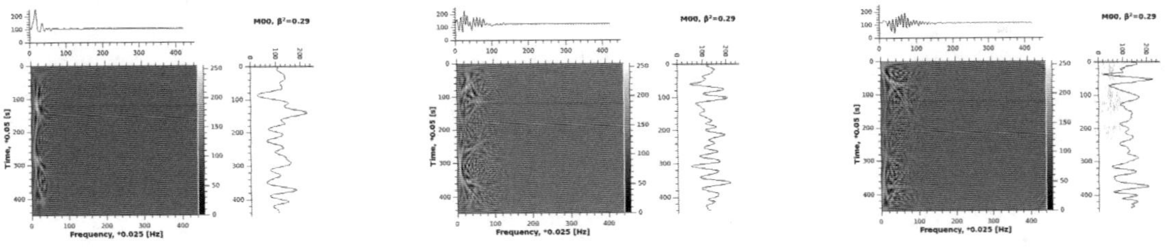

Figure 2. Time-frequency spectra for components of zero moment in conditions of low, average and hight turbulence for the 16x16 mm square aperture.

The work was supported by Russian Foundation for Basic Research. RFBR Project A-15-08-07484.

References

[1] Cohen L.,*Time Frequency Analysis: Theory and Applications*, Prentice Hall, 1994.

[2] E. A. Babanin, V. V. Kapranov, N. A. Suhareva, V. Y. Tugaenko, O. M. Vokhnik "Space-temporal stochastic characteristics of complex amplitude for the sounding vector optical beam", Proc. of PIERS 2017.

EPJ Web of Conferences **161**, 02004 (2017) DOI: 10.1051/epjconf/201716102004
PECS-2017

Nonlinear properties of optically dense medium excited by multifrequency laser radiation

Konstantin Barantsev*, *Andrey* Litvinov, and *Evgeniy* Popov

Peter the Great St. Petersburg Polytechnic University, Department of theoretical physics, 29 Polytechnicheskaia str., Saint-Petersburg, 195251 Russia

Abstract. This work is devoted to nonlinear effects in the atomic medium excited by multifrequency laser radiation. We have considered collective light scattering in the optically dense medium in conditions of coherent population trapping. Propagation of intermode correlation coefficient was calculated and it was found that correlation can increase in the medium. Light shifts of the coherent population trapping resonances and ways of their suppression were analyzed. Propagation of radiation in the medium with closed excitation scheme was investigated.

Optical properties of an atomic medium can sufficiently depend on parameters of the light which is scattered in this medium. This fact is the appearance of nonlinear effects. So, under the conditions of electromagnetically induced transparency the medium with the Λ-scheme of energy levels becomes transparent for the probe pulse on one of the transitions when the control field interacts with the second transition [1]. Such transparence appears due to destructive quantum interference between the two channels of excitation in the Λ-scheme, that is known as coherent population trapping (CPT) [2,3]. It is possible to control this effect (and therefore the transparency of the medium) by variation frequency detuning or intensities of the field.

However, in optically dense medium there is changing in the parameters of radiation due to absorption and scattering of the last one, and the atoms of deeper layers interacts with modified radiation. Especially, the intensity, spectrum, initial phase and statistics can change in the space. In the case of the radiation with several carrier frequencies, which is used for excitation the CPT resonance, the cross-correlation between frequency components can be modified. Such phenomena, when light scattering on a single atom depends on scattering on the other atoms, are called collective effects [4,5].

This work is devoted to theoretical investigation of the collective light scattering by optically dense atomic medium under conditions of the CPT resonance. Topicality is due to using of CPT in such areas as quantum frequency standards [3], magnetometry [6,7], quantum memory and deceleration of light [8,9], etc. Propagation of cross-correlations between the modes of the radiation is investigated in the optically dense gas of the alkali atoms. It was found that level of correlations can increase when the radiation passes through the medium.

* Corresponding author: kostmann@yandex.ru

We analyzed the light shifts of the CPT resonance with the view to reducing their sensitivity to intensity fluctuations. The intensity ratio of the frequency components, which provides the lowest sensitivity of the light shift to changing in the total intensity, was found in the optically dense medium.

Particular attention is drawn to the mediums with a closed excitation scheme excited by the laser radiation. Under these conditions the optical properties of the medium, especially CPT state, depend on initial phases of the radiation frequency components. It is possible both to destroy or restore CPT state by variation of the initial phases [10,11]. The paper [12] demonstrates the possibility of controlling absorption and refraction of the medium by means of changing in the initial phase of radiation.

In our work we calculate propagation of the radiation in the optically dense medium in the case of closed excitation scheme. We demonstrate that it is possible to control the space distribution of the absorption and refraction of the medium by changing in the initial phases of the frequency components of the radiation.

The work was supported by the Russian Science Foundation (project 17-12-01085).

References

[1] M. Fleischhauer, A. Imamoglu, J.P. Marangos, Rev. Mod. Phys. **77**, 633 (2005)
[2] E. Arimondo, Progress in optics **35**, 259 (1996)
[3] J. Vanier, Appl. Phys. **81**, 421, (2005)
[4] C.C. Kwong, T. Yang, D. Delande et al., Phys. Rev. Lett. **115**, 223601 (2015)
[5] V.M. Datsyuk, I.M. Sokolov, D.V. Kupriyanov et al., Phys. Rev. A **77**, 033823 (2008)
[6] M. Stahler, R. Wynands, S. Knappe et al., Opt. Lett. **27**, 1472 (2002)
[7] A. Akulshin, A. Celikov, V. Velichansky, Opt.Commun. **84**, 139 (1991)
[8] M.D. Lukin, Rev. Mod. Phys. **75**, 457 (2003)
[9] A. Imamoglu, J.P. Marangos, Rev. Mod. Phys. **77**, 633 (2005)
[10] D.V. Kosachiov, B.G. Matisov, Yu.V. Rozhdestvensky, J. Phys. B: At. Mol. Opt. Phys. **25**, 2473 (1992)
[11] S.J. Buckle, S.M. Barnett, P.L. Knight et al., Optica Acta **33**, 1129 (1986)
[12] H. Li, V.A. Sautenkov, Yu.V. Rostovtsev et al., Phys. Rev. A **80**, 023820 (2009)

EPJ Web of Conferences **161**, 02005 (2017) DOI: 10.1051/epjconf/201716102005
PECS-2017

Numerical calculations of the probabilities for quantum transitions in atoms and molecules by the path integral method

Alexander Biryukov*, and *Yana* Degtyareva

Samara National Research University, General and Theoretical Physics Department, Russian Federation

Abstract. The probabilities of molecular quantum transitions induced by electromagnetic field are expressed as path integrals of a real alternating functional. We propose a new method for computing these integrals by means of recurrence relations. We apply this approach to description of the two-photon Rabi oscillations.

The nonlinear interactions of microsystems with the laser radiation of various configurations and levels of intensity (an excitation and dissociation of molecules, ionization of atoms etc.) are under active study now. We consider it reasonable to investigate the probabilities of transitions in multilevel quantum systems nonperturbatively and without any restrictions on the structure of laser radiation, as in the intensity and in the form of pulses. To this end a method of expressing probabilities of quantum transitions as path integrals of the real alternating functionals given in the energy representation is under investigation [1-3].

It was proved [2-3] that the probability of a quantum transition from the system quantum state $|n_{in}\rangle$ at time instant t_0 to the state $|n_f\rangle$ for the time interval $(t_f - t_0)$ under the interaction with the electromagnetic field of a frequency Ω can be represented as a path integral of a real functional in the energy representation:

$$P(n_f,t_f \mid n_{in},t_0) = A\delta n_{K+1}, m_{K+1}\delta n_{in}, m_{in} \prod_{k=1}^{K} \mathbf{R}_k \cos\left(\sum_{k=1}^{K+1} \Delta S(n_k, m_k, n_{k-1}, m_{k-1}, \xi_{k-1}, \zeta_{k-1}) \right), \quad (1)$$

where

$$\mathbf{R}_k = \sum_{n_k=1}^{N} \sum_{m_k=1}^{N} \int_0^1 d\xi_k \int_0^1 d\zeta_k, \; k = 1,2,...$$

$$\Delta S(n_k, m_k, n_{k-1}, m_{k-1}, \xi_{k-1}, \zeta_{k-1}) = S(n_k, n_{k-1}, \xi_{k-1}) - S(m_k, m_{k-1}, \zeta_{k-1}),$$

$$S(n_k, n_{k-1}, \xi_{k-1}) = 2\pi(n_k - n_{k-1})\xi_{k-1} + \Omega^R_{n_k n_{k-1}} \cos(2\pi(n_k - n_{k-1})\xi_{k-1} - (\Omega - \omega_{n_k n_{k-1}})t_{k-1})\Delta t$$

is the action for the transition in the energy representation, n_i, m_i are the quantum numbers prescribing a state of the system, A is a constant that preserves the normalization condition

* Corresponding author: biryukov_1@mail.ru

DOI: 10.1051/epjconf/201716102005

$\sum_{n_f} P(n_f, t_f \mid n_{in}, 0) = 1$, $\Omega^R_{n_k n_{k-1}}$ is the Rabi frequency, $\omega_{n_k n_{k-1}}$ is a frequency of transitions between states $|n_k\rangle, |n_{k-1}\rangle$, $t_{k-1} = (k-1)\Delta t$.

The numerical calculation of probabilities according to (1) successfully describes the process for short intervals $(t_f - t_0)$, but leads to considerable errors for long time intervals $(t_f - t_0)$. In this work, we propose a method that makes it possible to calculate the probabilities of quantum transitions by formula (1) for any time interval using the recurrent representation method.

Let us consider functions

$$P_{\cos}(n_K, m_K, t; n_{in}, m_{in}, t_0) = \prod_{k=1}^{K-1} \mathbf{R}_k \cos\left(\sum_{k=1}^{K} \Delta S(n_k, m_k, n_{k-1}, m_{k-1}, \xi_{k-1}, \zeta_{k-1})\right),$$

$$P_{\sin}(n_K, m_K, t; n_{in}, m_{in}, t_0) = \prod_{k=1}^{K-1} \mathbf{R}_k \sin\left(\sum_{k=1}^{K} \Delta S(n_k, m_k, n_{k-1}, m_{k-1}, \xi_{k-1}, \zeta_{k-1})\right).$$

Then the quantum transition probability can be expressed as follows

$$P(n_f, t_f; n_{in}, t_0) = A\, \delta_{n_K = n_f, m_f} \delta_{n_{in} m_{in}} P_{\cos}(n_f, m_f, t; n_{in}, m_{in}, t_0).$$

The program for the numerical calculation of the integral (1) is based on the recurrence relations for the auxiliary probability functions $P_{\cos}$, $P_{\sin}$:

$$P_{\cos}(n_k, m_k, t_k; n_{in}, t_0) = \mathbf{R}_{k-1} \{\cos[\Delta S(n_k, m_k, n_{k-1}, m_{k-1}, \xi_{k-1}, \zeta_{k-1})] \times$$
$$\times P_{\cos}(n_{k-1}, m_{k-1}, t_{k-1}; n_{in}, t_0)\} -$$
$$- \mathbf{R}_{k-1} \{\sin[\Delta S(n_k, m_k, n_{k-1}, m_{k-1}, \xi_{k-1}, \zeta_{k-1})] P_{\sin}(n_{k-1}, m_{k-1}, t_{k-1}; n_{in}, t_0)\};$$

and

$$P_{\sin}(n_{k-1}, m_{k-1}, t_{k-1}; n_{in}, t_0) = \mathbf{R}_{k-2} \{\sin[\Delta S(n_{k-1}, m_{k-1}, n_{k-2}, m_{k-2}, \xi_{k-2}, \zeta_{k-2})] \times$$
$$\times P_{\cos}(n_{k-2}, m_{k-2}, t_{k-2}; n_{in}, t_0)\} +$$
$$+ \mathbf{R}_{k-2} \{\cos[\Delta S(n_{k-1}, m_{k-1}, n_{k-2}, m_{k-2}, \xi_{k-2}, \zeta_{k-2})] P_{\sin}(n_{k-2}, m_{k-2}, t_{k-2}; n_{in}, t_0)\}.$$

Given the initial conditions we can determine functions $P_{\cos}(n_k, m_k, t_k; n_{in}, t_0)$ and $P_{\sin}(n_k, m_k, t_k; n_{in}, t_0)$, and then we calculate the probability of the quantum transition from the state $|n_{in}\rangle$ at the instant of time $t = t_0$ to the state $|n_k\rangle$ at the instant $t_k > t_0$.

The method successfully describes a behavior of a number of quantum systems, and in particular, the experimentally observed two-photon oscillations of Rabi [4].

References

[1] G.V. Ryazanov, Sov. J. Exp. Theor. Phys. **35**, 1 (1958)
[2] A.A. Biryukov, M.A. Shleenkov, Teoreticheskaia fizika **13** (2012) (in Russian)
[3] A.A. Biryukov, M.A. Shleenkov, Samara State Technical Univ. **19**, 2 (2015)
[4] T.R. Gentile, B.J. Hughey, D. Kleppner, et al., Phys.Rev. A **40**, 9 (1989)

EPJ Web of Conferences **161**, 02006 (2017) DOI: 10.1051/epjconf/201716102006
PECS-2017

Quasistatic thermal and nonlinear processes of photoconversion of high-density optical radiation by multilayer structures

Arkadiy Blank[1,2,⋆], *Anton* Razuvaev[1], *Natalia* Suhareva[2], and *Vjatcheslav* Tugaenko[1]

[1] *S.P. Korolev Rocket and Space Corporation "Energia" (Russian Federation)*
[2] *M.V. Lomonosov Moscow State University (Russian Federation)*

Abstract. The results of the systematic experimental analysis of the thermal nonlinear electro-optic properties of photoelectric converters with silicon vertical cells in comparison with solar elements and elements on the basis of In/Ga/As are presented. The parameters of the linear and quadratic approximations for the investigated dependences are determined, that allows constructing a scalable analytic model of the converter with a given type of the working elements switching.

Perspective technologies of the directed delivery of powerful spatially concentrated continuous energy flows are associated with high-density photoelectric converters, ensuring much higher efficiency of photoelectric transformation than that of the Shock-Quiester threshold for single-layer systems. A typical example of such converters designed for flow densities over $10W/cm^2$ is a layered silicon structure offered by MH GoPower [1, 2]. Real operating conditions for such photo converters revealed a number of significant features and limitations, associated essentially with the thermal- and photo-self-action of the converted flow at high radiation concentrations and insufficient heat removal from the site. First of all, this is the thermal drift of the characteristics, the nonlinearity of the response to the changes in the radiation flow density, and the thermodynamic instability of the system.

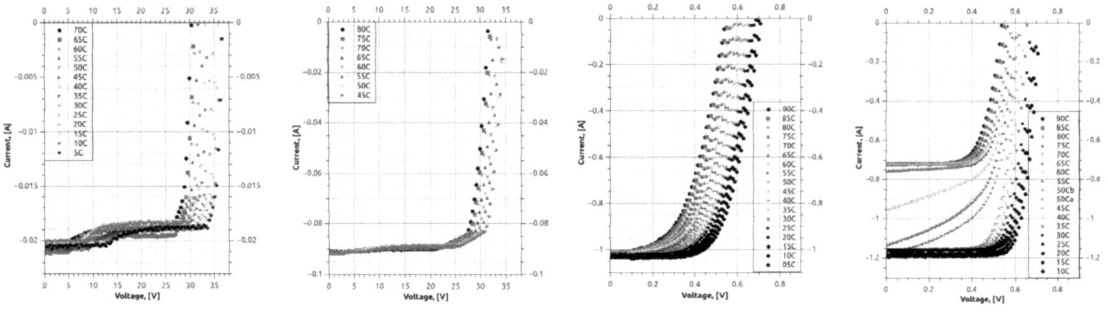

Figure 1. Example of the i-v characteristic for Si vertical cells (left and left-middle for various beam power), solar cell (right-middle) and InGaAs cell (right)

[⋆]e-mail: blankarkadiy@physics.msu.ru

 DOI: 10.1051/epjconf/201716102006

In the experiments we used the sources of single-mode and multimode continuous radiation having power up to 120 W and operating at wavelengths of 808 nm and 1082 nm. The choice of specific wavelengths reconciles the energy band structure of the silicon converter. The collimated beam covered the sample almost uniformly, the temperature stabilization of the sample surface was provided by the Peltier element, the spatial temperature profile of the site and the laser beam profile was fixated in the wavelength range from 12 μm to the visible range.

The control of the photoelectric conversion process is based on the analysis of the set of the i-v characteristics control values. They are considered to represent the sequence of the change in the photoconductor operating mode during the applied voltage change (Fig. 1). To measure the i-v characteristics a four-point scheme for different temperatures of the PVC surface, the photocurrent density, and the wavelength was used. Significant temperature drops in the range from -80^oC up to $+130^oC$, arising during the operation of space-based photo convertors, are accompanied by the drift of the boundaries of the spectral conversion band and the change in the carriers concentration. The reversed directivity of the characteristics is related to the position of the generation frequencies in the photo conversion band, since $\lambda = 1062nm$ lies on the red band boundary. Modules of the temperature coefficients dependence at the wavelength inside the range near to its edge ($+0,0007^oC-1$ and $-0,0009^oC - 1$) are close in magnitude within the measurement inaccuracy.

The maximum power produced by the FEP at the energy densities of the laser beam used in the temperature range from 0^oC to $+90^oC$ can be calculated from the semi-empirical equation of a quasi-stationary process using a linear approximation of the temperature dependence and quadratic dependence of the flow density of photo radiation:

$$W_{PVC} = \alpha_1(T)W_{LB} + \alpha_2(T)W_{LB}^2 \quad (1)$$

here W_{LB} – is laser beam power, W_{PVC} – is power generated by the solar cells, parameters α_1, α_2 are defined as follows:

$$\alpha_1(T) = 0.14 - 3.3 * 10^{-4} * T\ ,\ \alpha_2(T) = -7.7 * 10^{-4} + 2.5 * 10^{-5} * T. \quad (2)$$

Note that both parametric dependencies presented are sign-alternating. If one needs to estimate the partial contributions of the of multi-section FEP matrices having known temperature distribution and radiation flow density over the surface their using allows to solve a wide range of technological problems of selecting the optimal elements commutation for a given type of energy-carrying beams.

The work was supported by Russian Foundation for Basic Research. RFBR Project A-15-08-07484.

References

[1] Perales, Mico, et al. "Characterization of high performance silicon-based VMJ PV cells for laser power transmission applications.", Proc. of SPIE **9733**, (2016)

[2] Perales, Mico, et al. "Design and characterization of a novel power over fiber system integrating a high power diode laser.", Proc. of SPIE **10086**, (2017)

EPJ Web of Conferences **161**, 02007 (2017) DOI: 10.1051/epjconf/201716102007
PECS-2017

Wavelet-analysis of the non-stationary signal of fluorescence of an ensemble of point scatterers

Sergey Bozhokin*, and *Igor* Sokolov

Peter the Great St. Petersburg Polytechnic University, 195251 St. Petersburg, Russia

Abstract. On the basis of quantum microscopic approach we calculate dynamics of fluorescence of a cold atomic ensemble confined to special atomic trap and excited by pulse radiation. Using continuous wavelet transform with adaptive Morlet wavelet we analyze spectral composition of the fluorescence and its modification with time after the exciting pulse is switched off. Dependence of the spectrum on direction of fluorescence and its polarization is also analyzed.

The purpose of this work is a theoretical analysis of the dynamics of the secondary emission of atomic ensembles cooled to sub-Doppler temperatures in special traps and excited by pulsed radiation.

Atomic clouds in traps are well described by a model of motionless point scatterers and allow a consistent quantum microscopic description based on solution of the non-stationary Schrödinger equation [1-3] for a joint system consisting of atoms and an electromagnetic field. Based on this approach we calculate the time dependence of the field strength of an electromagnetic wave emitted by the considered atomic assemblies. The main attention is focused on the analysis of the spectral composition of this radiation and its changes with time after the excitation is swished off.

For the spectral analysis of non-stationary signals, a number of approaches are currently used. To select the best of them, we compared the two most effective - the Gabor window conversion and the continuous wavelet transform (CWT) [4-6]. This comparison was carried out on the basis of calculation of Pearson's correlation coefficients for model signals with known time and spectral properties. The comparison showed some advantages of the CWT method, especially when using the adaptive Morlet wavelet (AMW) [6] which contains a control parameter m that allows, if necessary, to increase either the temporal or the spectral resolution.

All specific calculations have been made on the basis of CWT using AMW for atomic clouds having a spherical shape with a random but uniform on average distribution of atoms. Averaging was carried out from this random distribution. The radius of the cloud was $R = 45\lambda\!\!\!^{-}$, the density of atoms n_0 was $n_0\lambda\!\!\!^{-3} = 0.002$ (here $\lambda\!\!\!^{-} = 1/k$, k is the wave number of the incident radiation). The exciting pulse was assumed to be resonant and circularly polarized, its duration τ_L was chosen much less than the lifetime τ_a of the excited states of the atoms: $\tau_L / \tau_a = 0.1$, which makes it possible to excite efficiently collective states in a

* Corresponding author: bsvjob@mail.ru

wide spectral interval. Fluorescence signals in different directions were calculated for helicity-preserving $H \parallel H$ and helicity-nonpreserving $H \perp H$ polarization channels.

For all the cases considered, a complex time dependence of the investigated signals is obtained. The fluorescence intensity decreases in accordance with a multiexponential law. For the short times after the end of the exciting pulse the damping rate exceeds the rate of spontaneous decay of free atoms. Superradiance is observed. For long times, the main contribution is made by long-lived collective states, and the sub-radiation effect takes place. When calculating the spectral composition under conditions of a significant attenuation of the fluorescence intensity, we analyze the signals normalized to the instantaneous value of their amplitude.

Fig. 1 illustrates the spectrum of radiation at $\theta = \pi/20$ (with respect to propagation direction of the incident wave) for one polarization channel. Fig. 1a shows the CWT $|V(\nu,t)|$ of the fluorescence signals, and Fig. 1b demonstrates a frequency $\nu_m(t)$, which corresponds to the maximal value $|V(\nu,t)|$ for a given instant of time t. Fig. 1a corresponds to m =50, at which the spectral resolution is well determined. In Fig. 1b the AMW control parameter is $m = 1$, and in this case the temporal modification of the spectra is determined in detail.

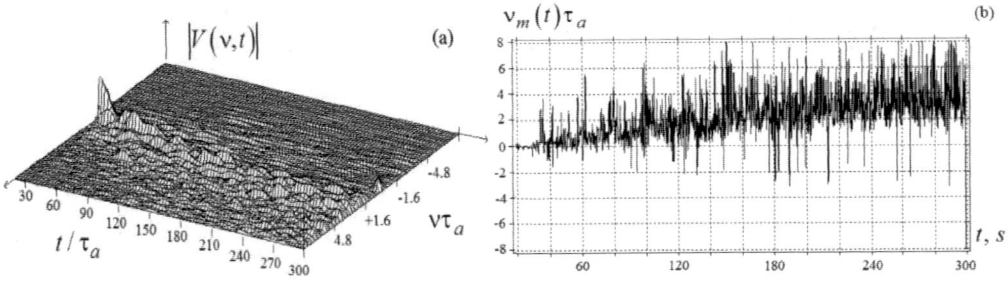

Fig. 1. CWT $|V(\nu,t)|$ and time dependence of frequency $\nu_m(t)$ for $H \perp H$, $\theta = \pi/20$.

Figs. 1a and 1b show essential difference between the emission spectra of a cloud and the spectra of free atoms, even in the case of a relatively dilute medium. These features of the spectrum are due to the presence of randomly formed pairs of closely located atoms. Such pairs form a kind of quasimolecule, the parameters of the excited states of which differ substantially from the parameters of atomic states by the energies and by the lifetimes. The contributions of these quasimolecules become most noticeable at large times $t \gg \tau_a$ after the end of the excitation, when the monatomic excitations have already completely relaxed.

Our calculations revealed also a significant difference among the spectra for different polarization channels as well as for different directions of fluorescence. These differences were explained by the features of the energetic spectra of the diatomic clusters.

This work was supported by the Russian Science Foundation (project No 17-12-01085).

References

[1] I.M. Sokolov, D.V. Kupriyanov, M.D. Havey, J. Exp. Theor. Phys. **112**, 246 (2011)
[2] Ya.A. Fofanov, A.S. Kuraptsev, I.M. Sokolov et al., Phys. Rev. A **87**, 063839 (2013)
[3] A.S. Kuraptsev, I.M. Sokolov, Phys. Rev. A **90**, 012511 (2014)
[4] S.A. Mallat, *Wavelet tour of Signal Processing* (NY, Academic Press, 2008)
[5] C.K. Chui, *An Introduction to Wavelet* (NY, Academic Press, 1992)
[6] S.V. Bozhokin, S.V. Zharko, N.V. Larionov et al., Tech. Phys. **62**, 837 (2017)

Three-dimensional few-circle optical pulses in the inhomogeneous environment of carbon nanotubes in an optical resonator

Ilya Dvuzhilov[1,*], *Yulia Dvuzhilova*[2], *Mikhail* Belonenko[1,2], *Irina* Zaporotskova[1], and *Natalia* Boroznina[1]

[1]Volgograd State University, Volgograd 400062, Russia
[2]Volgograd Institute of Business,Volgograd 400010, Russia

Abstract. We consider the task about few-circle optical pulses dynamics (light bullets) in the inhomogeneous environment of carbon nanotubes in a resonant optical cavity. Our calculations show that the optical resonator can create conditions, which can provide monitoring and controlling the shape and the velocity of the optical pulse. We show that the propagation of the light bullets is stable.

1 Introduction

Carbon nanotubes (CNTs) because of their nonlinear optical properties have been used as generator of a media with unique features. CNTs have height interest in the research community owing to the simplicity of their structure and their unique properties, which in turn contributed significantly to both the development in optical pulses propagation studies, as well as the development of optical devices based on them [1-2]. Ability to use CNTs as a medium for the formation of light bullets is the one of the most important feature.

Usually the propagation of optical pulses is considered in a uniform CNT environment that does not allow one to control the pulse velocity. However, if the propagation velocity of light bullets is determined by the refractive index of the medium and can vary within a wide range, one can perform a further modulation in the refractive index. In turn, this favours the formation of media with a modulated refractive index, thereby enabling the control of the propagation velocity of light bullets as well as the delay time.

2 Basic equations

The electromagnetic pulse propagation in carbon nanotube with optical resonator array has been considered, wherein both the pulse electric field and the current are directed along the nanotube axis.

For carbon nanotubes of the zigzag type, namely (m, 0), the dispersion relation for the energy of conduction electrons reads [3]:

* Corresponding author: dvuzhilov.ilya@volsu.ru

$$\varepsilon_s(p) = \pm\gamma\sqrt{1 + 4\cos(ap)\cos(\pi s / m) + 4\cos^2(\pi s / m)}\,, \qquad (1)$$

where $s = 1, 2 \ldots m$, nanotube has a type $(m,0)$, $\gamma \approx 2.7$ eV, $a = 3b / 2\hbar$, $b = 0.142$ nm - is a distance between neighboring carbon atoms.

Maxwell equations for non-magnetic dielectric environment bring to view:

$$\frac{\partial^2 \boldsymbol{E}}{\partial z^2} + \frac{\partial^2 \boldsymbol{E}}{\partial r^2} - \frac{1}{c^2}\frac{\partial^2 \boldsymbol{E}}{\partial t^2} + \frac{4\pi}{c}\frac{\partial j}{\partial t} + \frac{4\pi}{c^2}\frac{\partial^2 P}{\partial t^2} = 0, \qquad (2)$$

where **E** - is the electric field of the light wave, j - is the electron current density, t - is the time, and c- is the light velocity in the medium., $P=\alpha E$.

Solving the problem, we neglect the diffraction spreading of the laser beam in the direction along the CNT axis. We ignore the substrate electric field as well. Note that since both the typical size of CNTs and the distance between them are much smaller than the spatial domain typical size in which a few-cycle pulse is localized, we can use the approximation of continuous medium and consider the current distributed over the volume.

The performed investigation allows us to make the following conclusions:

The three-dimensional few-cycle optical pulse (light bullet) propagation in inhomogeneous medium of carbon nanotubes in optical resonator is stable.

At large times the pulse remains localized along the axis of the cylindrical resonator.

The formation of a stable structure, which occurs due to the reflection of waves from the walls of the resonator and subsequent interference. Such an interaction makes it possible to reduce the width of the pulse in the direction of the transverse axis of the resonator.

References

[1] A.V. Zhukov, R.E. Bouffanais, M.B. Belonenko, N.N. Konobeeva, Yu.V. Nevzorova, T.F. George, Eur. Phys. J. D **69**, 129 (2015)

[2] A.V. Zhukov, R.E. Bouffanais, G. Fedorov, M.B. Belonenko, J. Appl. Phys. **114**, 143106 (2013)

[3] A.V. Zhukov, R. Bouffanais, M.B. Belonenko, I.S. Dvuzhilov, Yu.V. Nevzorova, Appl. Phys. B **123**, 196 (2017)

Formation of solitons in a waveguide at second harmonic generation

Sergey Sazonov, *Alexey* Kalinovich*, *Irina* Zakharova, *Mariya* Komissarova, and *Pavel* Shestakov

Lomonosov Moscow State University, Faculty of Physics, Leninskie Gory, Moscow, 119991 Russia

Abstract. Second harmonic generation in a quadratic nonlinear waveguide is analyzed analytically and numerically. Formation of spatio-temporal solitons is demonstrated. They can form both positive and negative coefficients of group velocity dispersion due to the waveguide geometry. Analytical results are confirmed by computer simulations.

The propagation of a pulse-beam of optical radiation in a planar waveguide with cubic nonlinearity is accompanied by a space-time collapse, the nature of which depends on the input parameters. A medium with quadratic nonlinearity has long been used to avoid collapse [1] and to generate a stable spatio-temporal solitons.

In most theoretical works devoted to spatio-temporal solitons in quadratic nonlinearity, studies were conducted for a uniform medium [1-4]. In a waveguide of the same geometry in addition to the competition between nonlinearity, dispersion and diffraction, focusing properties of the waveguide play an important role [3, 5].

Pulsed second harmonic generation in a planar waveguide with a parabolic profile of the refractive index at the combined effect of diffraction and dispersion is described by the following system of equations for the complex amplitudes of the fundamental frequency ψ_1 and second harmonic ψ_2:

$$i\left[\frac{\partial\psi_1}{\partial z}+\left(\varepsilon_1\alpha_1\frac{x^2}{a_1^2}+\delta\right)\frac{\partial\psi_1}{\partial\tau}\right]+\frac{\beta_2^{(1)}}{2}\frac{\partial^2\psi_1}{\partial\tau^2}-\gamma_1\psi_1^*\psi_2 e^{i(2k_1-k_2)z}=\varepsilon_1 q_1\frac{x^2}{a_1^2}\psi_1+\frac{c}{2n_1\omega}\frac{\partial^2\psi_1}{\partial x^2}, \quad (1)$$

$$i\left[\frac{\partial\psi_2}{\partial z}+\left(\varepsilon_2\alpha_2\frac{x^2}{a_2^2}-\delta\right)\frac{\partial\psi_2}{\partial\tau}\right]+\frac{\beta_2^{(2)}}{2}\frac{\partial^2\psi_2}{\partial\tau^2}-\gamma_2\psi_1^2 e^{-i(2k_1-k_2)z}=2\varepsilon_2 q_2\frac{x^2}{a_2^2}\psi_2+\frac{c}{4n_2\omega}\frac{\partial^2\psi_2}{\partial x^2}. \quad (2)$$

Here, $n_{1,2}$ are the refractive indices for the first and second harmonics in the center of the waveguide, $\beta_2^{(1,2)}$ are the coefficients of group velocity dispersion, $\tau=t-\frac{1}{2}\left(\frac{1}{v_{g1}}+\frac{1}{v_{g2}}\right)z$,

$\gamma_1=\frac{4\pi\omega}{cn}\chi_2(-\omega;2\omega)$, $\gamma_2=\frac{4\pi\omega}{cn}\chi_2(\omega;\omega)$, $\delta=\frac{1}{2}\left(\frac{1}{v_{g1}}-\frac{1}{v_{g2}}\right)$, $\alpha_{1,2}=\frac{n_{1,2}^2-1}{2cn_{1,2}}$, $q_{1,2}=\omega\alpha_{1,2}$.

* Corresponding author: kalinovich@gmail.com

Possible modes of propagation of the pulse beams at the main and doubled frequencies depend on the competition between the focusing properties of nonlinearity and the focusing properties of the waveguide, as well as from the critical parameter, $\eta = 0{,}43\frac{n^2}{n^2-1}\frac{a^2 l_D}{l_{dis}^3}$,

where $l_{dis} = \frac{2\tau_0^2}{|\beta_2|}$ - the dispersion length, $l_D = \frac{n\omega}{c}R_0^2$ - the diffraction length. There are four main scenarios:

1) under defocusing nonlinearity ($\beta_2 > 0$) and defocusing waveguide ($\varepsilon_{1,2} = +1$) there are no stable modes of propagation;

2) under defocusing nonlinearity ($\beta_2 > 0$) and a focusing waveguide ($\varepsilon_{1,2} = -1$) the formation of solitons due to waveguide geometry is possible;

3) at focusing nonlinearity ($\beta_2 < 0$) and defocusing waveguide ($\varepsilon_{1,2} = +1$) a stable propagation is possible under the condition $\eta > 1$;

4) at focusing nonlinearity ($\beta_2 < 0$) and focusing waveguide ($\varepsilon_{1,2} = -1$) a stable propagation of solitons is always possible.

Fig. 1 shows a numerical solution of the system (1) – (2) close to the two-component soliton that is formed at the focusing nonlinearity and the focusing waveguide. The following parameters were used $\beta_2 = -0.01$, $\varepsilon_{1,2} = -1$, $\delta = 0$, $\alpha_{1,2} = 0$, $\gamma_1 = 1$, $\gamma_2 = 0.5$, $q_1 = q_2 = 1$, $\frac{c}{2n_1\omega} = 0.1$, $\frac{c}{4n_2\omega} = 0.05$. The incident pulse was taken in the form $\psi_1 = E_1 \mathrm{sech}^2(x)\mathrm{sech}^2(\tau)$, $E_1 = 10$, $\psi_2 = 0$. The second harmonic was generated in the process of radiation propagation in the waveguide. This resulted in an optical bullet spread on almost invariably.

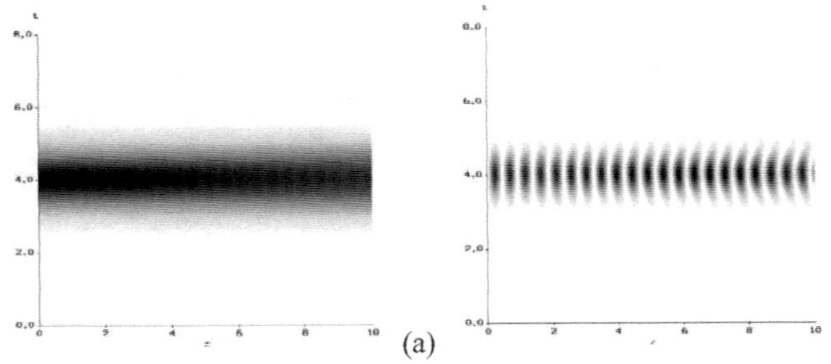
(a) (b)

Fig. 1. The intensity distribution in the cross sections tz demonstrates the formation and propagation of two-component soliton in a focusing waveguide with focusing nonlinearity. (a) fundamental harmonic, (b) second harmonic.

This work is supported by the Russian Science Foundation (Project No 17-11-01157).

References

[1] A.A. Kanashov and A.M. Rubenchik, Physica D **4**, 122 (1981)
[2] B.A. Malomed, P. Drummond, H. He et al., Phys. Rev. E **56**, 4725 (1997)
[3] S.V. Sazonov, Phys. of Wave Phenomena **24**, 31 (2016)
[4] S.V. Sazonov, J. Phys. Soc. Japan **85**, 124404 (2016)
[5] H. Guo, B. Zhou, M. Steinert et al., Opt. Lett. **40**, 629 (2015)

EPJ Web of Conferences **161**, 02010 (2017) DOI: 10.1051/epjconf/201716102010
PECS-2017

Electrically controlled phases of partially polarized light and orientational Kerr effect in liquid crystal ferroelectrics

Alexei D. Kiselev[1,⋆], *Vladimir V.* Kesaev[2,⋆⋆], and *Evgeny P.* Pozhidaev[2,⋆⋆⋆]

[1]*Saint-Petersburg National Research University of Information Technologies, Mechanics and Optics, 49 Kronverksky Pr., St. Petersburg 197101, Russia*
[2]*Lebedev Physical Institute RAS, 53 Leninskiy Ave., Moscow 119991, Russia*

Abstract. We study the electro-optic properties of subwavelength-pitch deformed-helix ferroelectric liquid crystals illuminated with partially polarized light. In an experimental setup based on the Mach-Zehnder interferometer, it is found that the interference pattern crucially depends on the degree of polarization of the incident light. We evaluate the electric field dependence of both the Pancharatnam relative phase and the geometric phase for the general case of nonunitarily evolving mixed polarization states.

Liquid crystal (LC) spatial light modulators have been widely employed to modulate amplitude, phase, or polarization of light waves in space and time. Ferroelectric liquid crystals (FLCs) represent most promising chiral liquid crystal material which is characterized by very fast response time. However, most of the FLC modes are not suitable for phase-only modulation devices because their optical axis sweeps in the plane of the cell substrate producing undesirable changes in the polarization state of the incident light. In order to get around the optical axis switching problem the system based on the orientational Kerr effect in a vertically aligned deformed helix ferroelectric LC (DHFLC) with subwavelength helix pitch was suggested in [1, 2]. In the geometry of a uniform lying helix, phase modulation of light in DHFLC cells was also studied using the experimental technique based on a Mach-Zehnder two-arm interferometer [3].

In our recent paper [4], we have found that, for unpolarized incident light, the electro-optic response is insensitive to the effect of electric-field-induced rotation of in-plane optical axes and can be described in terms of the electrically dependent Pancharatnam phase. In this paper, we examine the electro-optic behavior of the planar aligned DHFLC cells illuminated with partially polarized light and extend our theoretical analysis of the Pancharatnam and geometric phases to the general case of mixed polarization states that undergo nonunitary and non-cyclic evolution.

We consider a DHFLC layer of thickness D with the z axis normal to the bounding surfaces. The geometry of a uniform lying DHFLC helix with subwavelength pitch where the helix (twisting) axis $\hat{\mathbf{h}}$ is directed along the x axis will be our primary concern. According to [3, 5, 6], optical properties of such cells can be described by the effective dielectric tensor of a homogenized DHFLC helical structure. The electric-field-induced anisotropy is generally biaxial so that the dielectric tensor is characterized by the three generally different principal values (eigenvalues): $\epsilon_\pm = n_\pm^2$ and $\epsilon_z = n_z^2$.

⋆e-mail: alexei.d.kiselev@gmail.com
⋆⋆e-mail: vladimir.kesaev@gmail.com
⋆⋆⋆e-mail: epozhidaev@mail.ru

The in-plane principal optical axes (eigenvectors) $\hat{\mathbf{d}}_+ = \cos\psi_d\,\hat{\mathbf{x}} + \sin\psi_d\,\hat{\mathbf{y}}$, $\hat{\mathbf{d}}_- = \hat{\mathbf{z}}\times\hat{\mathbf{d}}_+$ are rotated about the vector of electric field, $\mathbf{E}\parallel\hat{\mathbf{z}}$, by the azimuthal angle ψ_d. In the low electric field region, the electric field dependence of the angle ψ_d is approximately linear: $\psi_d \propto E$, whereas the electrically induced part of the principal refractive indices, $n_\pm$ and n_z, is typically dominated by the Kerr-like nonlinear terms proportional to E^2 [3, 6]).

The input lightwave field is characterized by the polarization density matrix ρ_0 that, in the circular basis $\hat{\mathbf{e}}_\pm = (\hat{\mathbf{x}} \pm i\hat{\mathbf{y}})/\sqrt{2}$, can be written in the form: $2\rho_0 = \sigma_0 + P_0(\hat{\mathbf{s}}_0\cdot\boldsymbol{\sigma})$, $\sigma_0 = \mathrm{diag}(1,1)$, $\boldsymbol{\sigma} = (\sigma_1,\sigma_2,\sigma_3)$, where $\boldsymbol{\sigma}_i$ are the Pauli matrices; $\hat{\mathbf{s}}_0 = (\sin(2\theta_0)\cos(2\phi_0), \sin(2\theta_0)\sin(2\phi_0), \cos(2\theta_0))$ is the normalized Stokes vector and P_0 is the degree of polarization. Similarly, the transmission matrix of the DHFLC cell, $\mathbf{T}_s(h)$, where $h = k_{\mathrm{vac}}D = \omega/cD$ is the thickness parameter, can be expressed in terms of the vector $\hat{\mathbf{s}}_d = (\cos 2\psi_d, \sin 2\psi_d, 0)$ that can be regarded the normalized Stokes vector of a light beam linearly polarized along the optic axis $\hat{\mathbf{d}}_+$ and the transmission coefficients, t_+ and t_- [6].

We can now apply the interferometry based approach formulated in [7] to the case of plane wave which is normally incident on the DHFLC cell and is characterized by the polarization density matrix rho_0 and derive the expressions for the Pancharatnam and geometric phases, Φ_P and Φ_g, acquired by the light transmitted through the DHFLC cell. The result reads

$$\Phi_P = \Phi + \arg[\mathrm{Re}\,\tilde{F}_P + i\tilde{P}\,\mathrm{Im}\,\tilde{F}_P] = \arg\mathrm{Tr}[\mathbf{T}_s(h)\rho_0], \quad (1)$$

$$\tilde{F}_P = \cos\Delta\Phi\cos[\theta_\lambda(h) - \theta_\lambda(0)] + i\sin\Delta\Phi\cos[\theta_\lambda(h) + \theta_\lambda(0)], \quad (2)$$

$$\Phi_g = \arg[\mathrm{Re}\,\tilde{F}_g + i\tilde{P}\,\mathrm{Im}\,\tilde{F}_g],\ \tilde{F}_g = \tilde{F}_P e^{-i\tilde{\Phi}_d}, \quad (3)$$

$$\tilde{\Phi}_d = \int_0^h \cos(2\theta_\lambda(s))\Delta\Phi_s'(s)\mathrm{d}s,\ \cos(2\theta_\lambda) = q_3/|\mathbf{q}|,\ |\mathbf{q}| = \sqrt{q_3^2 + q_0^2[1 - (\hat{\mathbf{s}}_d\cdot\hat{\mathbf{s}}_0)^2]} \quad (4)$$

where $\Phi = (\Phi_+ + \Phi_-)/2$ is the averaged phase shift; $\Delta\Phi = (\Phi_+ - \Phi_-)/2$ is the phase retardation; $\tilde{P} = \dfrac{\sqrt{p_+(0)p_+(h)} - \sqrt{p_-(0)p_-(h)}}{\sqrt{p_+(0)p_+(h)} + \sqrt{p_-(0)p_-(h)}}$, $p_\mu(0) = \dfrac{1+\mu P_0}{2}$, $p_\mu(h) = \dfrac{\mathrm{Tr}(\rho) + \mu|\mathbf{q}|}{2}$, $q_0 = P_0[\tau_+^2 - \tau_-^2]$, $q_3 = 2\tau_-\tau_+ + P_0[\tau_+^2 + \tau_-^2](\hat{\mathbf{s}}_d\cdot\hat{\mathbf{s}}_0)$, $\mathrm{Tr}(\rho) = \tau_+^2 + \tau_-^2 + 2P_0\tau_+\tau_-(\hat{\mathbf{s}}_d\cdot\hat{\mathbf{s}}_0)$, $\tau_\pm = (|t_+| \pm |t_-|)/2$.

In conclusion, our analytical results show that, in contrast to the total Pancharatnam phase, the geometric phase is determined by the phase retardation and is independent of the averaged phase shift. In addition, the electric field dependence of the phases enters through the angle between the Stokes vectors $\hat{\mathbf{s}}_0$ and $\hat{\mathbf{s}}_d$ decribing the effect of electrically induced rotation of the optical axes. This effect is supressed in the limiting cases of unpolarized and circular polarized incident light. The effects of nonunitarity are governed by the difference between the transmittance coefficients: $|t_+| - |t_-|$.

We acknowledge financial support from RFBR under Grants 16-02-00441 a, 16-29-14012 ofi_m and 16-42-630773 p_a.

References

[1] E.P. Pozhidaev, A.D. Kiselev, A.K. Srivastava, V.G. Chigrinov, H.S. Kwok, M.V. Minchenko, Phys. Rev. E **87**, 052502 (2013)
[2] E.P. Pozhidaev, A.K. Srivastava, A.D. Kiselev, V.G. Chigrinov, V.V. Vashchenko, A.V. Krivoshey, M.V. Minchenko, H.S. Kwok, Optics Letters **39**, 2900 (2014)
[3] S.P. Kotova, S.A. Samagin, E.P. Pozhidaev, A.D. Kiselev, Phys. Rev. E **92**, 062502 (2015)
[4] V.V. Kesaev, A.D. Kiselev, E.P. Pozhidaev, Phys. Rev. E **95**, 032705 (2017)
[5] A.D. Kiselev, E.P. Pozhidaev, V.G. Chigrinov, H.S. Kwok, Phys. Rev. E **83**, 031703 (2011)
[6] A.D. Kiselev, V.G. Chigrinov, Phys. Rev. E **90**, 042504 (2014)
[7] D.M. Tong, E. Sjöqvist, L.C. Kwek, C.H. Oh, Phys. Rev. Lett. **93**, 080405 (2004)

Direct measuring of single-cycle mid-IR light bullets path length in LiF by the laser coloration method

Sergey Chekalin[1], *Victor* Kompanets[1,*], *Andrey* Kuznetsov[2], *Alexander* Dormidonov[3], and *Valerii* Kandidov[3]

[1]Institute of Spectroscopy RAS, Moscow, Troitsk, 108840 Russia
[2]Irkutsk Branch of Institute of Laser Physics SB RAS, Irkutsk, 664033 Russia
[3]Moscow Lomonosov State University, Physics Department and International Laser Center, Moscow, 119991 Russia

Abstract. A colour-centre structure formed in a LiF crystal under filamentation of a femtosecond mid-IR laser pulse with a power slightly exceeding the critical power for self-focusing has been experimentally and theoretically investigated. A single-cycle light bullet was recorded for the first time by observation of strictly periodic oscillations for the density of the color centers induced in an isotropic LiF crystal under filamentation of a laser beam with a wavelength tuned in the range from 2600 to 3900 nm, which is due to the periodic change in the light field amplitude in the light bullet formed under filamentation under propagation in dispersive medium. The light bullet path length was not more than one millimeter.

In contrast to the well-studied phenomena of the light beam squeezing due to self-focusing and wideband ultrashort pulse compression the formation of light bullets (LBs) with a high spatiotemporal light field localization is the result of simultaneous and matched laser radiation selfcompression both in space and time in nonlinear dispersive media. The numerical simulations suggest that such compression method may potentially deliver single optical cycle pulses in the 2–4 μm spectral range. In spite of existing methods for a single-cycle pulse duration measurements such LBs have not been recorded so far. The main reason is that LB diameter is about a few wavelengths and when we measure the whole beam not only the duration of the LB but also essentially longer duration of a background surrounding the LB is measured. The sensitivity of our recording technique is not sufficient to take a proper measurement with a lesser aperture. We proposed a novel method for such measuring based on a well known (from the end of the last century) phenomenon, specific only to near-single-cycle pulses, which is very important in attosecond physics - a cyclic transformation of the light field amplitude caused by the phase shift between the carrier wave and the wave packet envelope under propagation in a dispersive medium [1]. This transformation may be recorded by the laser coloration method (LCM), based on microscopic observation of laser-induced long-lived luminescent color centers (CC), which gives a unique possibility to register the filament spatial structure with an accuracy better than one micron after exposure

* Corresponding author: kompanetsvo@isan.troitsk.ru

and allows to restore the light field structure taking into account multiphoton character of the CC formation [2]. LiF is a unique material for such investigations because it has significantly brighter CC luminescence intensity in comparison with other alkali halide crystals that allowed us to observe a photoinduced transformation inscribed by a single LB produced by only one mid-IR exciting pulse. The CCs are created on a picosecond timescale, so they do not influence the femtosecond pulse dynamics. Opposite to on-line experiments, which demand many pulses exposure and where, in addition, the registration of supercontinuum radiation and plasma emission is inevitable, in LCM the structure of extremely compressed wave packet is recorded via appearance of the CCs in LiF after just single pulse passed through the material, and may be investigated much later by measuring the luminescence of CCs in the weak pump field. At last in LCM only the near-axial region of the beam in the filament without surrounding background is recorded, which has the smallest duration (that is just LB). These features distinguish LCM from the most advanced methods of LB dynamics study, such as 3D imaging technique, and allow us to answer the debated question about light bullet path length. It should be noted that on the base of experiments developed in the accumulation mode by averaging over a large series of LBs, many authors interpreted the extended light channels formed by mid-IR femtosecond pulses in transparent dielectrics as long-living LBs, the scale of which does not change along several centimeters distance. These results are contradictory to our numerical and experimental investigations [3] according to which the LB path length is much shorter.

In our LCM experiments the regular variation of the mid-IR LB parameters in LiF was revealed through the periodic CCs structure formed in a LiF crystal under filamentation of a single femtosecond mid-IR laser pulse with a power slightly exceeding the critical power for self-focusing [4]. Strictly periodic oscillations have been detected for the first time for the density of the CCs induced in an isotropic LiF crystal under filamentation of a laser beam with a wavelength tuned in the range from 2600 to 3900 nm. At the wavelengths shorter than 2500 nm such oscillations have not been observed. This experiment confirmed formation of near-single cycle LB in the presence of strong anomalous group velocity dispersion which 'breathing' in filament is a cyclic transformation of the light field amplitude and diameter caused by the phase shift between the carrier wave and the wave packet envelope in dispersive medium in agreement with our numerical modeling [5]. Such an experiment allowed us to unambiguously answer the debated question about light bullet path length which appeared to be of not more than one millimeter.

This work was supported by the Presidium of the Russian Academy of Sciences (Program "Extreme laser radiation: physics and fundamental applications").

References

[1] L. Xu, C. Spielmann, A. Poppe, T. Brabec, F. Krausz, and T.W. Hänsch, Opt. Lett. **21**, 2008 (1996)

[2] E.F. Martynovich, A.V. Kuznetsov, A.V. Kirpichnikov, E.V. Pestryakov, S.N. Bagayev, Quantum Electron. **43**, 463 (2013)

[3] E.O. Smetanina, V.O. Kompanets, A.E. Dormidonov, S.V. Chekalin, V.P. Kandidov, Laser Phys. Lett. **10**, 105401 (2013)

[4] A.V. Kuznetsov, V.O. Kompanets, A.E. Dormidonov, S.V. Chekalin, S.A Shlenov, V.P. Kandidov, Quantum Electron. **46**, 379 (2016)

[5] S.V. Chekalin, V.O. Kompanets, A.V. Kuznetsov, A.E. Dormidonov, V.P. Kandidov, Laser Phys. Lett. **13**, 065401 (2016)

DOI: 10.1051/epjconf/201716102012

Three-dimensional few-cycle optical pulses in germanene with damping and amplification

Natalia Konobeeva[1,*], and *Mikhail* Belonenko[1,2]

[1]Volgograd State University, Volgograd, 400062 Russia
[2]Volgograd Institute of Business, Volgograd, 400048 Russia

Abstract. We investigate three-dimensional few-cycle optical pulse propagation in germanene. We obtain an effective equation for description of the multidimensional pulses dynamics with damping and amplification. We study the effects which are observed with the change in the initial pulse shape at the entrance to the sample.

In recent years, the graphene-like structures generated increased interest from researchers because of the unique properties valuable in many practical applications. Such materials include, for example, germanene. It is a single layer structure constructed of germanium atoms, which are located at the sites of the hexagonal lattice [1]. This material was predicted by scientists in 2009, and only in 2014 it was synthesized. The main distinguishing feature of germanene from other graphene-like materials is a strong spin-orbit interaction, which is three orders of magnitude greater than the spin-orbit interaction in graphene. Therefore, it leads to a significant impact on the few-cycle pulse propagation in this medium.

Let us write the dispersion law for germenene [2]:

$$\varepsilon_{\sigma\xi} = \pm\sqrt{v^2k^2 + \frac{1}{4}\left(\Delta_z - \sigma\xi\Delta_{SO}\right)^2} \tag{1}$$

where σ is the electron spin, ν is the velocity of Dirac electrons, $\mathbf{k}=(k_x, k_y)$ is the quasi-momentum of the electrons, Δ_{SO} is the spin-orbit interaction, Δ_z is the potential on the one site, $\Delta_z=E_zd$, d is the distance between the sublattice planes, σ_i, τ_i are the Pauli matrices [3].

Using similar arguments as in [4], we can write the equation for the pulse propagation in cylindrical coordinate system as follows:

$$A_{tt} - \Gamma A_t = \frac{1}{r}\frac{\partial}{\partial r}\left(r\frac{\partial A}{\partial r}\right) + \frac{\partial^2 A}{\partial z^2} + \frac{1}{r^2}\frac{\partial^2 A}{\partial \phi^2} + 4\pi j(A)exp\left(-t/t_{rel}\right) \tag{2}$$

here the current density j is calculated using the electron spectrum (1), A is the vector-potential, parameter Γ describes the pumping of the electric field and, consequently, its amplification, c=1 is the light velocity. Here, the relaxation of electrons with time t_{rel} is phenomenologically taken into account, which leads to a decay of the current.

Equation (2) was solved numerically using classical numerical scheme. The initial condition in the three-dimensional case was chosen as a very short pulse, consisting of one, and two oscillations (n=1 and 2 respectively). The evolution of the electromagnetic field upon

* Corresponding author: yana_nn@volsu.ru

its spreading over a sample is shown in Fig. 1. The dependence of the pulse shape on the different initial velocity is presented in Fig. 2.

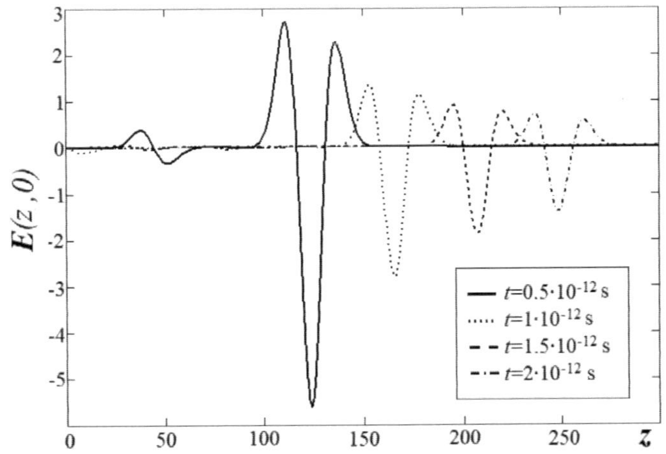

Fig. 1. The electric field at different instances of time at the r=0, n=1.

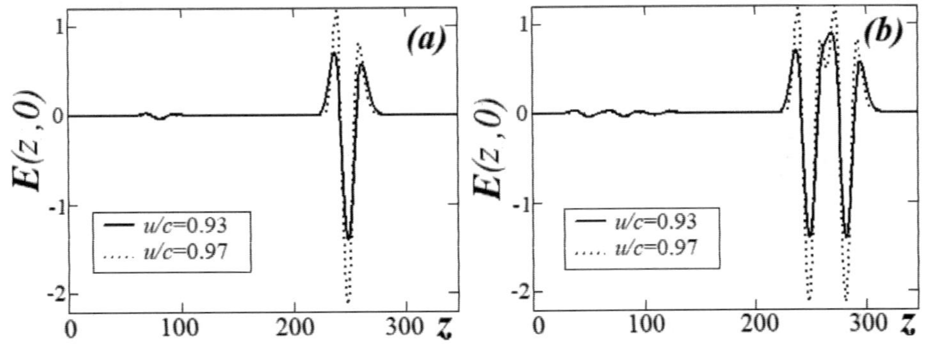

Fig. 2. The electric field at various values of pulse velocity: a) n=1; b) n=2.

A three-dimensional few-cycle optical pulse propagates with a damped amplitude and with the shape conservation, which is due to the balance of losses of its energy and the energy scattering caused by dissipation. The velocity of the pulse at the entrance to the sample has a strong effect on its evolution.

This work was supported by the Ministry of Education and Sciences of the Russian Federation (project no. MK-4562.2016.2, project no. 2.852.2017/4.6).

References

[1] S. Lebeque, O. Eriksson, Phys. Rev. B **79**, 115409 (2009)
[2] A.V. Zhukov, R. Bouffanais, N.N. Konobeeva, M.B. Belonenko, Phys. Lett. A **380**, 3117 (2016)
[3] M. Ezawa, J. Phys. Soc. Jpn. **84**, 121003 (2015)
[4] A.V. Zhukov, R. Bouffanais, E.G. Fedorov, M.B. Belonenko, J. Appl. Phys. **114**, 143106 (2013)

 DOI: 10.1051/epjconf/201716102013

Ground-state level spin systems within Bernasconi model

A.N. Leukhin*, V.I. Bezrodnyi, A.A. Voronin, and N.A. Kokovikhina

Mari State University, 1 Lenin sq., Yoshkar-Ola, 424000 Russia

Abstract. The design problem for law autocorrelation binary sequences (LABS) is a notoriously difficult computational problem which is numbered as the problem number 005 in CSPLib. In statistical physics LABS problem can be interpreted as the energy of N interacting Ising spins. This is a Bernasconi model for a one N-dimensional chain of quantum particles. Ground-state level quantum system within the frame of Bernasconi model is binary sequence with lowest level of peak-sidelobes of aperiodic autocorrelation function. The new results of an exhaustive search for optimal binary sequences with minimum peak sidelobe up to length N=86 are demonstrated. The Bernasconi model exhibits features of a glass transition like a jump in the specific heat and slow dynamics and aging. The Bernasconi model the high temperature phase of Ising spin system reproduces exactly approximation.

Let us consider the one-dimensional spin lattice $\mathbf{S}=(s_0,s_1,...,s_{N-1})$ consisting of N particles with an equal distances between neighbouring particles. Within one dimensional Ising model [1] the spin can assume one of two values $s_n=\pm1$ in each lattice node with a number n. The energy of such quantum system consists of pairwise of exchanged interaction between neighbouring atom spins and the interaction of the spins and an external magnetic field:

$$E(\mathbf{S})=-\sum_{n,\tau}J_{n\tau}s_ns_\tau-\sum_{n}h_ns_\tau \tag{1}$$

where indices n and τ denote the number of lattice nodes, h_n is the strength of the external magnetic field in the n-th node, and $J_{n\tau}$ is the energy of interaction for spins located in nodes n and τ.

In [2], Bernasconi pointed out the connection between the problem of searching for low energy states of a simplified the one dimensional Ising model (2) and that of constructing binary sequences $\mathbf{S}=(s_0,s_1,...,s_{N-1})$ of N length with low levels of autocorrelation.

There are two criteria for the optimality of binary sequences with low levels of aperiodic autocorrelation. The first is the minimax criterion, in which the maximum level of side lobe PSL

* Corresponding author: leukhinan@list.ru

$$PSL(\mathbf{C}) = \max_{1 \le \tau \le N-1} |C_\tau| \qquad (2)$$

must be minimal:

$$MPS = \min_{\mathbf{S}} PSL \,. \qquad (3)$$

The second is the MF (merit factor) coefficient characterizing the ratio of the main frame energy to that of the side lobes of the aperiodic autocorrelation function:

$$MF(\mathbf{C}) = \frac{N^2}{2\sum_{\tau=1}^{N-1} [C_\tau]^2} \,. \qquad (4)$$

Table 1 shows the constructed binary sequences with the length $N = 86$ and $PSL = 5$ with the maximum values of the coefficient MF.

Table 1. Optimal binary sequences with low level of side lobes

Binary sequence	*MF*
00001001111011010111101101001110110000010010011100110010000001100001 11001010101010000	7,125
00000000100010001101110101001100101100110001111011000010110110101011 0 0001110101111 0000	7,017
000000000001011010100101010011011100010101110011001010011011000100111 10010110011110000	6,522
000000001110000010000111101111011010100010101010011101011010011001010 01110010011011000	6,522
000000000001011010100101010011011100010101110011001010011011000100111 10010110011110000	6,522
000000000001110011100001010011100110011001101100100100101101010101010 11110000011110000	6,431
000000000100101011010110001001110101010100011000111100110001001101001 01100100111110000	6,431
000001001011001110011110110000001010001001010010101000001011110100110 00101110011111000	6,431
000000000001110011100001010011100110011001101100100100101101010101010 11110000011110000	6,431

This work was supported by grants from the RFBR (project No. 15-07-99514), the Ministry of Education and Science of the Russian Federation (project No. 2.2226.2017/Project and project No 2.9140.2017/Base).

References

[1] E. Ising, Z. Phys. **31**, 253 (1925)
[2] J. Bernasconi, J. Phys. **48**, 559 (1987)

EPJ Web of Conferences **161**, 02014 (2017) DOI: 10.1051/epjconf/201716102014
PECS-2017

Nonlinear optical response of a 2D quantum dot supercrystal

Ramil Malikov[1,*], *Igor* Ryzhov[2], and *Victor* Malyshev[3]

[1]M. Akmullah State Pedagogical University of Bashkortostan, 450000 Ufa, Russia
[2]A.I. Hertsen State Pedagogical University of Russia, 191186 St.-Petersburg, Russia
[3]Institute for Advanced Materials, University of Groningen, 9747 AG Groningen, The Netherlands

Abstract. We investigate theoretically the nonlinear optical response of a two-dimensional supercrystal comprized of semiconductor quantum dots. An isolated quantum dot is modeled as a three-level ladder-like system with ground, one-exciton, and biexciton states. It is shown that the optical response of supercrystal demonstrate a rich nonlinear dynamics, including bistability, self-oscillations, and dynamical chaos.

Supercrystals (SCs) comprising semiconductor quantum dots (SQDs) represent a class of new materials not existing in nature. Modern nanotechnology has in its disposal a variety of methods to design such systems [1]. Optical properties of SCs depend on the SQD's size, shape, and chemical composition, as well as on the lattice geometry and can be easily controlled [2].

We conduct a theoretical study of the optical response of a two-dimensional SC. Due to a high density of SQDs in the SC, the SQD-SQD dipole-dipole interaction plays an important role in the SC's optical response, both linear and nonlinear. This interaction provides a positive feedback which, together with the SQD's nonlinearity, results in a rich optical dynamics of SC, including bistability, self-oscillations, and dynamical chaos.

It is assumed that the SC undergoes an action of an external field of an amplitude E_0 and frequency ω_0, which is quasiresonant with the SQD allowed transitions. A single SQD is modelled as a three-level ladder-like quantum system with the ground $|1\rangle$, one-exciton $|2\rangle$, and bi-exction $|3\rangle$ states. Only the optical transitions $|1\rangle\leftrightarrow|2\rangle$ and $|2\rangle\leftrightarrow|3\rangle$ are dipole-allowed. They are characterized by the transition dipole moments d_{21} and d_{32}, transition frequencies ω_{21} and ω_{32}, and the spontaneous decay constants γ_{21} and γ_{32}. The frequency ω_{32} is down-shifted with respect to ω_{21} by an amount Δ_B being the biexciton binding energy.

The optical dynamics of an isolated SQD is governed by 3x3 density matrix. The field acting on a given SQD in the SC represents a sum of the external field and the field produced by all others SQDs in place of the given one. In this way, the total (retarded) dipole-dipole interaction is taken into account. As the mean dipole moment of an SQD depends on how strong the SQD is excited, the SQD-SQD interaction appears to depend on the current SQD's condition as well. Similar to a one-dimensional case [3], the near-zone part of the retarded SQD-SQD dipole-dipole interaction gives rise to a dynamic shift of the transition frequencies

* Corresponding author: rfmalikov@mail.ru

ω_{21} and ω_{32}, depending on the population difference of corresponding transitions. The SQD-SQD interaction in the far-zone stipulates a collective radiative relaxation of SQDs, also depending on the population difference. These two effects are described by the constants Δ_L (shift) and γ_R (relaxation). Furthermore, for a SC an inequality $\Delta_L >> \gamma_R$. is always fulfilled. Just these parameters translate into a positive feedback, leading to an unusual nonlinear SC's optical dynamics.

In Fig. 1, we show the results of calculations of the field magnitude $|E|$ inside the SC obtained under adiabatically sweeping the external field magnitude E_0 up and down for two conditions of excitation: the external field is in resonance (i) with the one-exciton transition, $\omega_0 = \omega_{21}$ (left panel) and (ii) with the coherent two-photon transition, $\omega_0 = \omega_{31}/2$ (right panel). As follows from Fig. 1, the system, similarly to the two-level model [3,4], demonstrates hysteresis: a sudden switching from one stable state to the other one, occurring at different magnitudes of the external field E_0. In other words, it is bistable. Second, depending on the excitation conditions, ether the upper branch of the hysteresis loop ($\omega_0 = \omega_{21}$) or the lower one ($\omega_0 = \omega_{31}/2$) shows windows of instability. From the analysis of the phase and Poincare maps of the field, we found that the character of these instabilities is ether of self-oscillation or chaotic nature. The inserts illustrate this.

The reflected from and transmitted through the SC fields show the identical dynamics. Furthermore, on the lower branch of hysteresis loop, the SC almost totally reflects the incident field, whereas on the upper branch, the reflection is negligible, so the system is almost transparent.

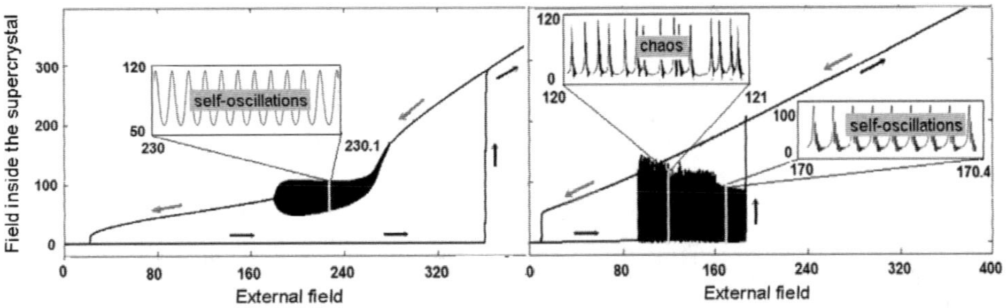

Fig. 1. The field magnitude $d_{21}|E|/\hbar$ inside the SC versus the external field magnitude $d_{21}E_0/\hbar$, obtained under adiabatically sweeping E_0 up and down (shown by arrows). Left panel - $\omega_0 = \omega_{21}$, $\Delta_B = 120\gamma_{21}$. Right panel - $\omega_0 = \omega_{31}/2$, $\Delta_B = 50\gamma_{21}$. The rest of parameters are: $\gamma_R = 100\gamma_{21}$, $\Delta_L = 1000\gamma_{21}$. The inserts show zoomed fragments of the dynamics.

On the basis our study, we believe that supercrystals comprising SQDs might serve, first, as an all-optical switch, second, as a generator of trains of ultrashort optical pulses (in the self-oscillation regime), and, finally, as a noise generator (in the chaotic regime). These features might be of interest for all-optical technologies and nanophotonics.

R. F. M. acknowledges the M. Akmullah State Pedagogical University of Bashkortostan for a support.

References

[1] M.P. Boneschanscher, W.H. Evers, J.J. Geuchies et al., Science **344**, 1377 (2014)
[2] A.S. Baimuratov, I.D. Rukhlenko, V.K. Turkov et al., Sci. Rep. **3**, 1927 (2013)
[3] V. Malyshev, P. Moreno, Phys. Rev. A **53**, 416 (1996)
[4] R.F. Malikov, V.A. Malyshev, Opt. Spectrosc. **122**, 955 (2017)

Capabilities of modal type liquid crystal spatial modulators for formation of light fields with orbital angular momentum

Svetlana Kotova*, *Aleksandra* Mayorova, and *Sergey* Samagin

Lebedev Physical Institute, Samara Branch, 221 Novo-Sadovaya Str., Samara, 443011 Russia

Abstract. The capabilities of modal LC modulator (LC focusator) for optical vortices formation are demonstrated. Two options of changing of the contact electrodes geometry providing a voltage jump and accordingly the jump for profile of phase transmission of modulator are proposed. The examples of distributions of intensity and phase for generated fields are presented.

Light fields with complex spatial structure and orbital angular momentum are very promising for contemporary optical technologies. Fields like that are useful for optical diagnostics, information processing, laser diagnostics, optical manipulation etc. [1]. The range of problems that can be solved using complex light fields, is constantly expanding [2]. Thus, the realization and improvement of methods for formation of such fields and for dynamic control of their characteristics is important. At the same time, for practical applications inexpensive, technologically simple devices working in a wide spectral region and characterized by relatively low light loss are of interest. Liquid crystal (LC) light modulators of modal type, and, in particular, LC focusator developed at Samara Branch of Lebedev Physical Institute can be used as such a device. Modal type modulators include in their design a high-resistance layer performing the function of the control electrode. The phase delay profile is generated by changing the voltage distribution applied to the electrodes as well as the reorientation of molecules of LC layer between the glass substrates covered with high-resistance and low-resistance coatings.

The aim of this work is to demonstrate the enhanced functionality of the LC focusator to generate light beams with angular orbital momentum by means of changing the contact electrodes geometry. Previously it was shown that the LC focusator allows one to focus light field into the point spot or line, and also to generate transverse intensity distributions in the form of various contours, while the sizes, positions and orientations can be controlled with the applied voltages [3-5]. But all fields generated with the LC focusator are characterized by zero orbital angular momentum, which limits their application.

To generate beams with orbital angular momentum (optical vortices) it is necessary to ensure the phase jump. Two ways of changing the configuration of the contact electrodes are proposed. In the first case, the contact electrode on one of substrates is divided by half, and this region is extended down to the middle of the aperture. And in the second case, two narrow

* Corresponding author: kotova@fian.smr.ru

linear electrodes are applied from the centre to the edge in the middle of the free side of one of the substrates. In both cases we have two additional closely spaced electrodes and we can realize a voltage jump and, therefore, the jump in the profile of phase delay applying to those electrodes the selected slightly different in magnitude potentials. The result is generation of angular orbital momentum for the light passing through LC modulator or reflected from it.

The results of simulation showed that in the first case it is possible to implement the phase delay like to the spiral axicon and to form the transverse intensity distribution in the shape of a ring at some distance from the LC modulator. To ensure more homogeneous phase distribution and intensity of the generated fields it is preferable to use the dual-frequency control mode. The control of voltage and phase delay distributions and respectively of the generated fields can be achieved by changing the potentials on one of the frequencies. The intensity and phase distributions of the light field generated by means of the LC modulator with dual-frequency control mode at a distance of 25 cm from it are presented as example in Fig. 1 (a, b).

The proposed configuration of modal LC focusator with additional narrow electrodes can be used as a spiral plate, which is clearly illustrated by the examples of intensity and phase distributions of the generated light fields in the far zone of diffraction shown in Fig. 1 (c, d). The field was calculated in the focal plane of the lens with a focal length of 1 m for the device with aperture 2 x 2 mm, limited by a diaphragm with a diameter of 1 mm.

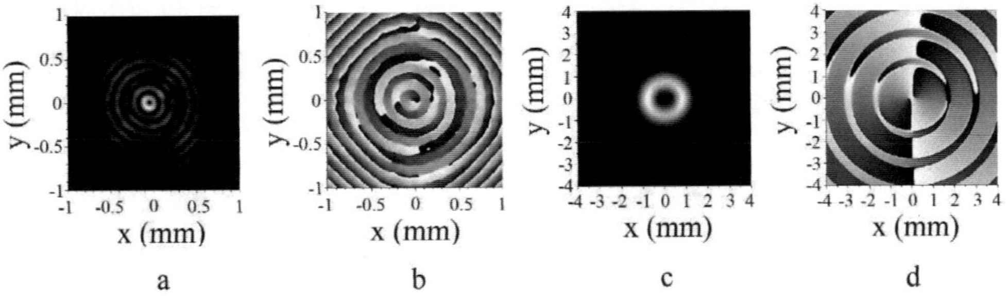

Fig. 1. The distributions of intensity (a, c) and phase (b, d) of light fields generated by LC focusator. a, b – The simulation results for spiral LC modulator with dual-frequency control mode at a distance of 25 cm from it. c, d - The simulation results for the LC focusator with additional electrodes at the focal plane of the lens.

The LC focusator is a technologically simple and relatively low-cost device. Besides, the use of solid electrodes allows us to form a smooth continuous profile of the phase delay and to control it smoothly. Due to these features in our opinion the suggested configurations of electrodes can be useful for the optical vortexes formation in optical manipulations tasks.

This research is financially supported by the Russian Foundation for Basic Research (projects No. 16-02-01055 and No. 16-42-630773).

References

[1] M. Padgett, Opt. Express **25**, 11265 (2017)
[2] H. Rubinsztein-Dunlop et al., J. Opt. **19**, 013001 (2017)
[3] A. Korobtsov, S. Kotova, N. Losevsky et al., J. Opt. **16**, 035704 (2014)
[4] S. Kotova, A. Mayorova, S. Samagin, J. Opt. **17**, 055602 (2015)
[5] S. Kotova, A. Mayorova, S. Samagin, J. Opt. Technol. **84**, 323 (2017)

Amplification of surface plasmon polariton wave in single-walled carbon nanotube using electric current pump

Sergey Moiseev[1,2,*], *Yuliya* Dadoenkova[1,3] , *Aleksey* Kadochkin[1], and *Igor* Zolotovskii[1]

[1]Ulyanovsk State University, Ulyanovsk, 432017 Russia
[2]Kotelnikov Institute of Radio Engineering and Electronics of the Russian Academy of Sciences, Ulyanovsk Branch, Ulyanovsk, 432011 Russia
[3]Donetsk Institute for Physics & Technology, Donetsk, 83114, Ukraine

Abstract. We propose a surface plasmon polariton amplification technique based on direct energy transfer from a dc electric current flowing in a carbon nanotube. It is shown that when the synchronization conditions are satisfied, when the surface plasmon polariton wave velocity is close to the drift velocity of the charge carriers in the nanotube, the surface wave is significantly enhanced.

1 Introduction and description of the system

Surface waves propagating along the boundary between two media, one of which has a negative dielectric permittivity in the whole frequency range, are called surface plasmon polaritons (SPPs). The SPPs are characterized by high field localization and penetrate in each adjacent medium at a depth of the wavelength order, decaying exponentially with the distance from the interface. The investigation of SPPs is of great interest due to wide possibilities of their practical application. However, the use of materials with negative permittivity (as a rule, conductive media) inevitably leads to ohmic losses.

In this paper, we propose a new mechanism of the SPP amplification due to the transfer of electromagnetic energy from an external source (drift current wave) into a surface wave. In this case a necessary condition of the interaction of these waves is synchronization condition, when the phase velocity of the surface wave approaches the charge drift velocity. The main idea of this approach is described in details in our paper [1], devoted to amplification of a SPP wave in a semiconductor film with graphene layer. In this paper, the proposed approach is applied to a single-walled carbon nanotube (CNT).

The CNT is assumed to be an infinitely thin cylindrical shell of the length L and radius $a = 5$ nm ($L >> a$) suspended in vacuum with voltage applied to its ends. We consider non-radiative TM-mode with non-zero component E_z of electric field along the CNT. The presence of a radial component of the electric field is necessary to ensure the interaction of SPP wave with an electric pump current flowing along the CNT. The field distribution of SPP and spectral characteristics of CNT are derived within hydrodynamic model for electron

* Corresponding author: serg-moiseev@yandex.ru

gas in CNT and using Maxwell equations and general boundary conditions for electromagnetic field [2].

2 Results of the numerical calculations

The equation describing the interaction of the electromagnetic wave and the electron current I_d writes:

$$\frac{dE_z}{dz} + i\frac{\omega}{V_{ph}}E_z = -\frac{1}{2}\left(\frac{\omega}{V_{ph}}\right)^2 2\frac{I_d |E_{0z}|^2}{\beta^2 P}, \tag{1}$$

where $V_{ph} = \omega/\beta$, ω and β are phase velocity, angular frequency and propagation constant of SPP, and P is power carried by SPP wave. Under influence of the SPP field the amplitude of the current is modulated along the CNT, and the current perturbation $\Delta I(x) = I(x) - I_d$ (I_d is unperturbed current) can be described by the equation:

$$\frac{d^2\Delta I}{dz^2} + 2i\frac{\omega}{V_0}\frac{d\Delta I}{dz} - \frac{1}{V_0^2}\left(\omega^2 - \omega_q^2\right)\Delta I = i\frac{\omega}{V_0}\frac{I_d}{2U_0}E_z, \tag{2}$$

where U_0 is potential difference, V_0 is charge carriers velocity in CNT, ω_q is reduced plasma frequency. Assuming that ΔI and E_z change along the CNT proportionally to $\exp(-iGz)$, G is the harmonic perturbation wavenumber, on the basis of a joint solution of Eqs. (1) and (2), we calculate numerically the SPP amplification coefficient $\alpha = \mathrm{Im}(G)$.

The synchronism condition is fulfilled at frequency $\omega_0 = 3 \cdot 10^{13}$ rad/s (see the intersection of the solid blue and dashed green lines in Fig. 1(a)) corresponding to infrared regime. Within the frequency interval $\Delta\omega \approx 8 \cdot 10^{13}$ rad/s the SPP amplification coefficient α (Fig. 1(b)) is about 10^6 cm^{-1} which is much larger than the SPP loss coefficient.

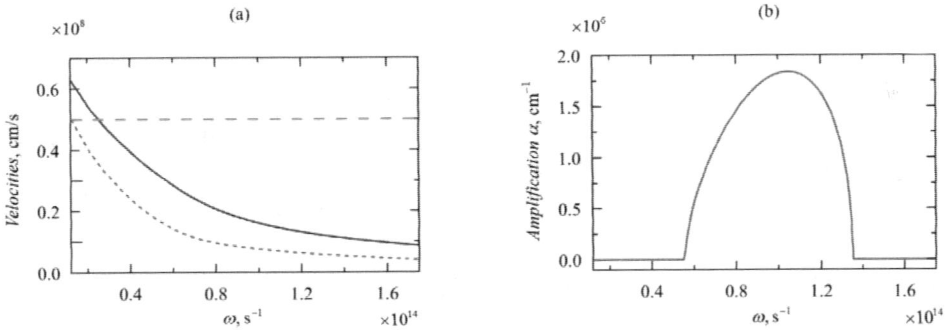

Fig. 1. Frequency dependences of: (a) SPP phase velocity (solid blue line) and group velocity (dotted red line), and drift current velocity $V_0 = 0.5 \cdot 10^8$ cm/s (dashed green line) in CNT; (b) SPP amplification coefficient α. The surface concentration of free electrons in CNT is $n_s = 10^{12}$ cm^{-2}.

This work was supported by the Ministry of Education and Science of the Russian Federation (Project No. 14.Z50.31.0015, State Contracts No. 3.7614.2017/P220, No. 3.5698.2017/P220, and No. 3.3889.2017), and the Russian Foundation for Basic Research (Project No. 17-02-01382).

References

[1] Y.S. Dadoenkova, S.G. Moiseev, A.S. Abramov, A.S. Kadochkin, A.A. Fotiadi, I.O. Zolotovskii, Ann. Phys. **529**, 1700037 (2017)

[2] A. Moradi, Phys. Plasmas **21**, 032106 (2014)

Modifications of laser correlation spectrometer for investigation of biological fluids

Elina Nepomnyashchaya[*], *Zoja* Zabalueva, *Elena* Velichko, and *Evgenii* Aksenov

Peter the Great St. Petersburg Polytechnic University, Institute of Physics, Nanotechnology and Telecommunications, 29 Polytechnicheskaya Str., St. Petersburg, 195251 Russia

Abstract. We suggest three modifications of a laser correlation spectrometer for analysis of complex multicomponent biological solutions based on: (i) dynamic light scattering with special signal processing for analysis of polydisperse solutions, (ii) use of polarimetric technique for shape analysis, (iii) cross-correlation processing. The schemes of the devices based on polarization and cross-correlation analysis are described.

1 Introduction

In modern medical physics the problem of research of biological solutions is solved by using various techniques, such as mass spectrometry, electronic and atomic microscopy, fluorescent spectroscopy [1] and nuclear magnetic resonance. Laser correlation spectroscopy (LCS) is used to study sizes of nanoparticles in liquids, however its applications for the analysis of multicomponent environments, such as blood serum, are limited. LCS is based on registration and analysis of light scattered from nanoparticles. We offer a combination of several registration schemes, which will allow distinguishing molecules in biological liquids, such as saliva and blood serum.

2 Experimental setup

The scheme of the experimental setup is presented in Fig. 1. A single-mode laser module with a power up to 2.5 mW and 650 nm wavelength was used. To achieve focusing, a nonspherical short-focus lens allowing us to obtain a 50 μm laser beam diameter was used, the calculated extent of caustic in the solution was 5 mm. Scattered radiation was registered under angles 5° — 175° by a single-mode optical fiber connected to a photoelectronic multiplier. For the 90° scattering angle, the calculated length of dispersion was 4.8 mm. The signal from the photoelectronic multiplier was digitized by an ADC module with frequencies of 100 kHz — 50 MHz and processed by a computer. This scheme allowed characterization of spherical nanoparticles with a high accuracy (± 0.5 nm).

To analyse aspheric nanoparticles, the polarimetric technique was used [2]. The scattered radiation was registered in two perpendicular polarizations, translational and rotational diffusion coefficients were calculated. Rotational diffusion coefficient gave us information

[*] Corresponding author: elina.nep@gmail.com

about shapes of the nanoparticles, which cannot be obtained with conventional LCS. By analyzing the polarization and depolarization scattering components simultaneously, we were able to determine the degree of nonsphericity for the albumin protein at different pH of the medium, which allowed us to determine real dimensions of the nonspherical molecules.

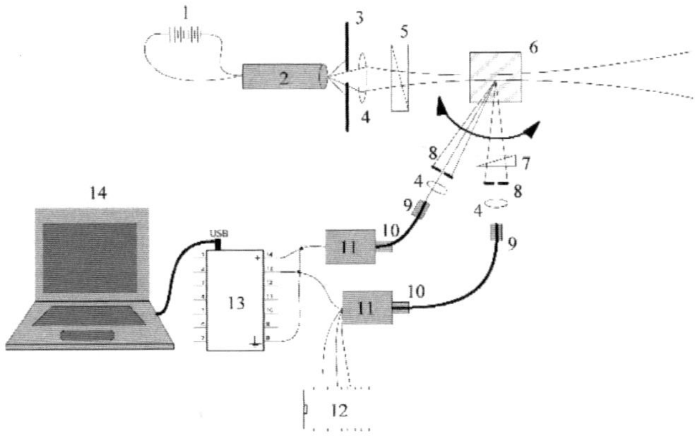

Fig. 1. Laser correlation spectrometer. 1 — rechargeable battery, 2 — laser, 3 — screen, 4 — lenses, 5 — polarizer, 6 — cuvette with the studied solution, 7 — output polarizer-analyzer, 8 — pinhole, 9,10 — input and output connectors of the single-mode fiber, 11 — photoelectronic multiplier, 12 — stabilized power supply, 13 — ADC module, 14 — personal computer.

Our scheme gave us a possibility of a simultaneous registration of scattering at different angles and allowed us to exclude contributions of multiple scattering and to analyse solutions with high particle concentrations (non-Gaussian statistics) [3]. This allows studies of such liquids as blood and saliva in an undiluted form.

The signal registered in all schemes of measurements represents a dependence of the intensity of scattered radiation on time. To identify regularities in the scattered signals, the autocorrelation or cross-correlation (in the case of a simultaneous registration of two scattered signals) functions were calculated. Further analysis was carried out by using the original modification of the Tikhonov's mathematical technique for ill-posed problems [4].

Such biological liquids as blood serum and plasma, saliva, solutions of complement proteins, albumin, thrombin, various amino acids and also magnetic liquids were investigated [5]. Polydisperse size distributions of nanoparticles in the biological fluids were obtained. Polarization LCS allowed analysis of the nonsphericity of nanoparticles. Our cross-correlation spectroscopic scheme will allow one to obtain these parameters for biological solutions in initial concentrations without the need for additional dilution and filtration, which can simplify the medical application of the device developed. The technique demonstrated a high efficiency in the analysis of biological solutions.

References

[1] S. Palmer, K. Litvinova, A. Dunaev, Biomed. Opt. Exp. 7, 1193 (2017)

[2] S.A. Dolgushin, I.K. Yudin, V.A. Deshabo, Biomed. Eng. **4,** 394 (2016)

[3] C. Urban, P. Schurtenberger, Progr. Colloid Polim. Sci. **110**, 61 (1998)

[4] E. Nepomnyashchaya, E. Velichko, E. Aksenov, J. Phys: Conf. Ser. **769,** 012025 (2016)

[5] E. Nepomnyashchaya, E. Savchenko, E. Velichko, E. Aksenov, J. of Biomed. Phot. and Eng. **2**, 040309 (2017)

 DOI: 10.1051/epjconf/201716102018

Light bullets in an Bragg environment with carbon nanotubes in the external magnetic field

Yulia Nevzorova[1], *Ilya* Dvuzhilov[1,*], *Sergey* Boroznin[1], *Irina* Zaporotskova[1], and *Mikhail* Belonenko[1,2]

[1]Volgograd State University, 100 University Prospect, Volgograd 400062, Russia
[2]Volgograd Institute of Business, 63 Kachintsev Str., Volgograd 400010, Russia

Abstract. On the basis of the solution of the collisionless Boltzmann equation for electrons in the conduction band of semiconductor carbon nanotubes, an effective equation for the dynamics of two-dimensional few cycle light pulses (light bullets) is obtained, in the case when the environment into which carbon nanotubes are placed has a spatially modulated refractive index. The distortion of the pulse shape in the presence of a magnetic field is obtained.

1 Introduction

Earlier, a number of papers [1, 2] studied the propagation of one-, two- and three-dimensional ultrashort optical pulses in a Bragg medium with carbon nanotubes and revealed the fact of their steady propagation. It was also found that in the case where the constant magnetic field is parallel to the axis of the carbon nanotube, the shape of the pulse can change and it becomes possible to control. As shown in [3] in this case, if the same electron-radio spectrum can be found exactly and in an arbitrarily strong magnetic field, which makes it possible to analyse the differences due to the presence and absence of a constant magnetic field.

2 Basic equations

As a rule, the study of carbon nanotubes' electronic structure is carried out in framework of analysis of dynamics of π-electrons in strong-coupling approximation. Consider an alternating electric field that is parallel to a constant magnetic field propagating in the CNT system. The Hamiltonian for electron system can be written in the form:

$$H = \sum_{ps} \varepsilon_s \left(p - \frac{e}{c} A(t) \right) a^+_{ps} a_{ps} \tag{1}$$

* Corresponding author: dvuzhilov.ilya@volsu.ru

where a^+_{ps}, a_{ps} are creation and annihilation operators of electrons at site p with spin s, is a hopping integral, determined by the electron wave function overlapping at adjacent sites. $A(t)$ - is the light pulse electric field, $\varepsilon_s(p)$ - electron dispersion law.

For carbon nanotubes of the zig-zag type, the dispersion law for electrons in the presence of a magnetic field parallel to the nanotube axis is:

$$\varepsilon_s\left(k_x,k_y,H\right)=\pm\gamma\sqrt{1+4\cos\left(ap_z\frac{3ak_z}{2}\right)\cos\left(\frac{\sqrt{3}ak_x}{2}\right)+4\cos^2\left(\frac{\sqrt{3}ak_x}{2}\right)} \tag{2}$$

where $\gamma \approx 2{,}7$ eV, ka belongs to the interval [- π; π], k is the wave vector along the CNT axis, a = 3b/2ħ, b = 0.152 nm are the distances between adjacent carbon atoms, k_z - wave vector along the nanotube axis.

$k_x=\frac{2\pi}{\sqrt{3}aM_0}\left(s+\frac{\Phi}{\Phi_0}\right)$ Φ – magnetic flux through the cross section of the nanotube, $\Phi_0=\frac{\hbar c}{e}$, $s = 1, 2, \ldots, M_0$, M_0 – the number of hexagons along the perimeter of the nanotube.

Using the standard Coulomb gauge, the pulse electromagnetic field can be described by the Maxwell equations as follows:

$$\frac{\partial^2 A_z}{\partial x^2}+\frac{\partial^2 A_z}{\partial y^2}-\frac{n^2(x,y)}{c^2}\frac{\partial^2 A_z}{\partial t^2}+\frac{q}{\pi\hbar}\sum_m c_m\sin\left(\frac{maq}{c}A_z(t)\right)=0 \tag{3}$$

$$c_m=\sum_m a_{ms}b_{ms}\quad b_{ms}=\int_{-q_0}^{q_0}dp_z\cos(map_z)F_0(\mathbf{p}) \tag{4}$$

where t is time, c is speed of light in the medium, $n(x; y)$ is the coefficient determining spatial variation of the carbon nanotube density, $F_0(\mathbf{p})$ - equilibrium Fermi distribution function.

Solving the problem, we neglect the diffraction spreading of the laser beam in direction along the CNT axis. We ignore the substrate electric field as well. Note that since both the typical size of CNTs and the distance between them are much smaller than the spatial domain typical size in which a few-cycle pulse is localized, we can use the continuous medium approximation and consider the current distributed over the volume.

It is established that a stable propagation of two-dimensional light bullets is possible in the presence of an external constant magnetic field directed along the nanotube axis in a Bragg medium with carbon nanotubes. Under the influence of a magnetic field, the shape of the pulse is distorted, the increase of the external magnetic field to a significant change in the propagation dynamics of a two-dimensional ultimately short optical pulse.

References

[1] M.B. Belonenko, Yu.V. Nevzorova, Bull. RAS: Physics **78**, 1619 (2014)

[2] A.V. Zhukov, R. Bouffanais, M.B. Belonenko, N.N. Konobeeva, Yu.V. Nevzorova, T.F. George, Eur. Phys. J. D **69**, 129 (2015)

[3] M.B. Belonenko, Yu.V. Nevzorova, E.N. Galkina, Bull. RAS: Physics **80**, 837 (2016)

DOI: 10.1051/epjconf/201716102019

Optical pumping of the atom magnetic moment under spin polarization equal to zero

Evgeniy Popov[1,*], *Sergey* Voskobojnikov[2], *Konstantin* Barantsev[1], and *Andrey* Litvinov[1]

[1]Peter The Great St. Petersburg Polytechnic University, Department of Quantum electronics, 29 Polytechnicheskaia str., Saint-Petersburg, 195251 Russia

[2]Peter The Great St. Petersburg Polytechnic University, High school of software engineering, 29 Polytechnicheskaia str., Saint-Petersburg, 195251 Russia

Abstract. Possibility of preparing of an alkali-atom's state with non-zero magnetic moment and zero spin-polarization was demonstrated in this work. There are zero average values of Pauli spin operators for all coordinates in the case. We suggest using a biharmonic laser field for preparing such states. Moreover both components of the laser field should be set to resonance with two sublines in D1 line of the alkali and have circular polarization. The amplitudes of the components needed to prepare the observed states was found. It is found interesting that polarization of the components of the laser may be orthogonal or parallel for preparing the states.

Generally, the spin polarization of alkali atoms is prepared with a laser field which has a circular polarization. In the case when a dark-state appears in the ground level the interaction between an atom in the dark-state and the laser field stops. Physically the dark-state is the Zeeman sublevel with the maximum value of the projection component of the atom's full angular momentum to the quantization axis. During the decay progress the atoms accumulate in this Zeeman sublevel. This leads to appearance of the non-zero expectation value for the atom spin operator or non-zero value of the spin-polarization. The atoms in this state have a magnetic moment which is determined by the gyromagnetic ratio of an electron and Lande factor of the energy level.

Preparing of the alkali atoms in the state with spin polarization is widely used in magnetometers, inertial navigation, experimental researches of spin-exchange interaction between atoms and in other [1,2].

The goal of the work was observation of spin polarization pumping with two laser fields, which are set to resonance with two transitions in D1 line of the alkali atom. Scheme of pumping is demonstrated in Fig. 1. Notice, we made transition from the hyperfine-levels basis to the basis of electron and nuclear states because the relaxation of an electron-state and a nuclear-state passes with different velocities under collisions with the buffer. This detail led to essential amplification of the mathematical model in comparison with the usual semi-classical model. An assembly of an alkali was described with density matrix $\hat{\sigma}$, which involved all sublevels in the D1 line of the alkali and all transitions between this

* Corresponding author: enp-tion@yandex.ru

sublevels. Dynamics of atoms in semi-classical model was determined by the Liouville equation [3]:

$$\frac{d\hat{\sigma}}{dt} = \frac{i}{\hbar}[\hat{\sigma}, \hat{H}] + \hat{\Gamma}_{rel}\{\hat{\sigma}\}, \quad (1)$$

where $\hat{H}$ is Hamiltonian, $\hat{\Gamma}_{rel}$ is relaxation superoperator which describes all processes of decay. We worked in dipole approximation:

$$\hat{H} = \hat{H}_0 - (\hat{\mathbf{d}} \cdot \mathbf{E}), \quad (2)$$

where $\hat{H}_0$ is Hamiltonian without laser field, $\hat{\mathbf{d}}$ is an operator of atomic dipole, $\mathbf{E}$ is a vector of the electric field which contains both components of the laser pumping:

$$\mathbf{E} = E_1\mathbf{l}_1 e^{-i\omega_1 t} + E_2\mathbf{l}_2 e^{-i\omega_2 t} + c.c.. \quad (3)$$

In the equation (3) E_n, $\mathbf{l}_n$ and ω_n are an amplitude, a polarization vector and a frequency of the n-th component of the laser pumping.

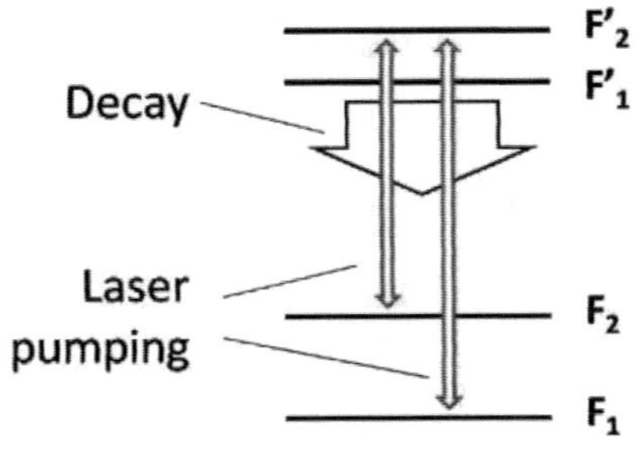

Fig. 1. Scheme of laser pumping of alkali's D1 line with biharmonic laser field

Further two regimes of laser pumping were observed: laser fields had the same polarizations $(\mathbf{l}_1\mathbf{l}_2^*) = 1$ and it had the orthogonal polarizations $(\mathbf{l}_1\mathbf{l}_2^*) = 0$. Variating amplitudes of the laser fields we found that spin polarization of the alkali became zero for certain values of the amplitudes. The most interesting fact is the effect was found for both regimes of pumping. We expected this effect for the second regime of the pumping, but reduction of spin polarization to zero for the same polarizations of laser fields surprised us.

In addition, we demonstrated that prepared to the state with zero spin polarization assembly of the alkali had non-zero expected value of the magnetic moment. In the case we couldn't use the Bloch's vectors for describing spin polarization of the alkali because a link between magnetization and spin polarization was not determined with an effective gyromagnetic ratio. We supposed that the considered states would have unusual magnetic properties and may be used in magnetometry, inertial navigation, medical physics.

The work supported by Russian Foundation for Basic Research (project No.16-32-00587)

References

[1] A.K. Vershovsky, A.S. Pazgalev, J. Tech. Phys. **76**, 108 (2006)
[2] J.C. Fang, J. Qin, Sensors **12,** 6331 (2012)
[3] W. Happer, Review of Modern Physics. **44**, 170 (1972)

EPJ Web of Conferences **161**, 02020 (2017) DOI: 10.1051/epjconf/201716102020
PECS-2017

The method of surface plasmon-polariton pulses generation via cooperative effects in a waveguide spaser

Alexander Shesterikov[1], *Mikhail* Gubin[1], and *Alexey* Prokhorov[1,*]

[1]Vladimir State University named after Alexander and Nikolay Stoletovs, Vladimir, 600000 Russia

Abstract. Cooperative effects arising under conditions of a 0-D model for a dense ensemble of semiconductor quantum dots located in a dielectric layer near a flat metal surface are considered. The threshold conditions of the effect are determined and the values of the complex refractive index of the dielectric are chosen, which make it possible to increase the efficiency of the formation of surface plasmon-polariton (SPP) pulses.

In this paper we consider the problem of SPP pulse generation during the collective exciton decay of excited quantum dots (QDs) with concentration N placed in a dielectric layer near a metal surface (Fig. 1a). The QD size is determined by condition of plasmon-exciton resonance (Fig. 1b), and the QD concentration corresponds to compensation of losses at the considering scales. The development dynamics of collective effects is described by self-consistent equations for population imbalance n_{21}, QD polarization ρ_{21}, and Rabi frequency Ω_0:

$$\dot{\rho}_{12} = -0.5\cdot\Gamma_\varepsilon\rho_{12} + \left(i\Omega_0^* + i\xi_0 u_R\rho_{12} + \xi_0 u_I\rho_{12}\right)n_{21}, \quad (1a)$$

$$\dot{n}_{21} = 2i\left(\Omega_0\rho_{12} - \Omega_0^*\rho_{21}\right) - 4\xi_0 u_I\left|\rho_{12}\right|^2 - \Gamma_\varepsilon\left(1+n_{21}\right),\ \dot{\Omega}_0 = -i\cdot t_R^{-2}\cdot\rho_{12} - \gamma_p\Omega_0, \quad (1b,c)$$

where $t_R = \dfrac{1}{g\sqrt{N_a}} = \sqrt{2\hbar\varepsilon_d\varepsilon_0 \dfrac{\partial S_n(\overline{\omega})}{\partial\overline{\omega}} / \left(S_n(\overline{\omega})\mu_{12}^2 N\right)}$ determines the characteristic time for the establishment of quantum correlations between QDs and depends on dipole moment of the interband transition μ_{12} and the Bergman's parameter $S_n(\overline{\omega}) = \mathrm{Re}(\varepsilon_d/(\varepsilon_d - \varepsilon_m(\overline{\omega})))$ for the metal-dielectric interface with dielectric permittivities of dielectric ε_d and metal $\varepsilon_m(\overline{\omega}) = 1 - \omega_p^2/\left(\overline{\omega}^2 + i\gamma_p\overline{\omega}\right)$, accounting the electron collision frequency γ_p and the plasmon frequency $\overline{\omega} = \omega_p/\sqrt{2}$, ω_p is the plasma frequency in metal; $g = \Omega_0/\varepsilon$ is the coupling constatnt, $\varepsilon = \sqrt{N_p}$, where N_p is the number of plasmons in volume V. The parameter Γ_ε in (1a) and (1b) determines spontaneous rate of losses in QD and

* Corresponding author: avprokhorov33@mail.ru

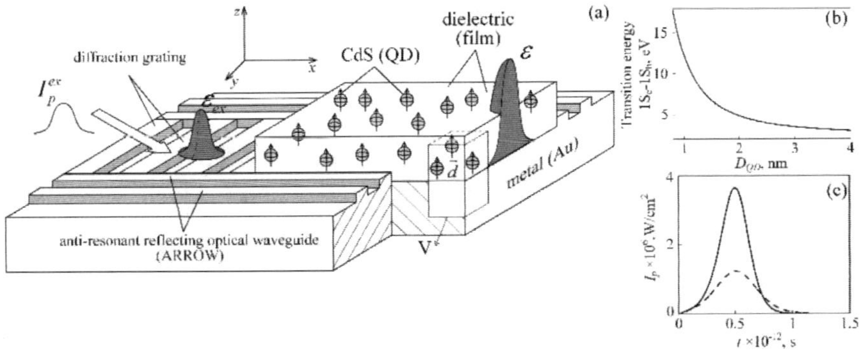

Fig. 1. (a) Formation scheme of SPP pulses in a planar metal/dielectric waveguide [1] pumped by CdS QDs [2]; (b) dependence of the transition energy on the CdS QD size; (c) intensity profiles of SPP pulses (1c) under condition $N = 2.83 \cdot 10^{22}$ m^{-3} for QDs with $D_{QD} = 1.56$ nm near a gold surface for $\Gamma_\varepsilon = 6.3 \cdot 10^{11}$ s^{-1} (dashed curve) and $\Gamma_\varepsilon = 0$ s^{-1} (solid curve) decay rates in QD.

$\xi_0 = N\mu_{12}^2 / 3\hbar\varepsilon_0$ is associated with accounting the local field of dielectric. The dispersion u_R and dissipation u_I corrections correspond to the case of dense ensemble of QDs doped in dielectric with refractive index $n = n_R + in_I$.

The threshold conditions under which QD concentration proves to be sufficient for the development of collective effects in a waveguide spaser even taking into account fast decay of the SPP in the metal were determined. The comparison of different regimes of SPP pulse formation in a metal/dielectric interface was carried out by use of simulation (1). The optimum values of n, which make it possible to completely compensate the relaxation processes in QD (Fig. 1c) were found. Using mean field approximation and assuming $t_R << 1/\gamma_p$ the self-consistent problem (1) was reduced to the nonlinear pendulum equation with additional term, which appears due to accounting u_I in (1a) and (1b) and provides fast coherent energy exchange between ensemble of QDs and SPP pulse. We also demonstrate the possibility of a strong spectral broadening and generation almost rectangular spectrum SPP pulses due to the existence of a strongly nonlinear near-field effects in the dielectric layer with a dense ensemble of chromophores. However, the possibility of a SPP supercontinuum generation in the system under study requires a substantial reduction in the pulse duration and the use of novel models [3] beyond the slowly-varying-envelope approximation. The possibility of control of SPP pulse dynamic by analogous an effect of triggered superradiance [4], as well as by means of an external magnetic field using magnetic nanostructures [5] is discussed.

The work was supported by RFBR (17-42-330001) and the Ministry of Education and Science of the Russian Federation in the framework of the state task VlSU 2017 in the field of scientific research.

References

[1] A.V. Shesterikov, M.Yu. Gubin, M.G. Gladush, A.V. Prokhorov, JETP **124**, 18 (2017)
[2] K.A. Magaryan et al., J. Lumin. **169**, 799 (2016)
[3] H. Leblond, P. Grelu, D. Mihalache, Phys. Rev. A **90**, 053816 (2014)
[4] P.V. Zinoviev, V.A. Zuikov, A.A. Kalachev et al., Las. Phys. **11**, 1307 (2001)
[5] P. Lupo, Z. Haghshenasfard, M.G. Cottam et al., Phys. Rev. B **94**, 214431 (2016)

EPJ Web of Conferences **161**, 02021 (2017) DOI: 10.1051/epjconf/201716102021
PECS-2017

Superradiance without inversion in a cavity: Effects of cavity losses and relaxation in the system of emitters

Igor Ryzhov[1,*], *Nikolai* Vasil'ev[1], and *Victor* Malyshev[2]

[1]A.I. Hertsen State Pedagogical University of Russia, St.-Petersburg, 191186 Russia
[2]Zernike Institute for Advanced Materials, University of Groningen, 9747 AG Groningen, The Netherlands

Abstract. We investigate theoretically inversionless superradiance of an ensemble of three-level emitters (with Λ-scheme of operating transition), placed in a cyclic cavity. Effects of cavity losses and emitter's relaxation on the dynamics of superradiance are studied.

The standard Dicke superradiance (SR) requires an initial population inversion of the operating transition to occur [1]. In the case of multilevel emitters (in particular, three-level atoms with the Λ-scheme of operating transitions considered here), this limitation is not crucial anymore: SR is possible even if the initial population of the upper level is smaller than the population of the doublet (SR without inversion, SRWI) [2-4]. In Ref. [2], it has been found that SRWI of an assembly of three-level Λ-emitters in a high-Q cavity shows a rich optical dynamics (from regular to chaotic), depending on the population of the upper level α and splitting of the doublet ω_{21}. These results have been obtained for a Hamiltonian system, i.e. neglecting the cavity losses and relaxation of coherence in the system of emitters. In reality, however, these factors are always present.

In this paper, we investigate theoretically the influence of relaxation on the regimes of SRWI found in Ref. [2] (see Ref. [4b] for details). In the Hamiltonian case and at a small doublet splitting, the SRWI dynamics represents a comb of chaotically repeating pulses with a repetition time on the order of the period of oscillations of the low-frequency coherence, $2\pi/\omega_{21}$ (see Fig. 1a). Just this regime of SRWI is the goal of our study.

1. Cavity losses. This channel of SR relaxation is characterized by the SR lifetime in the cavity T_c. Figure 1b shows an example of the SRWI dynamics for $T_c = 50\Omega^{-1}$, where Ω is the cooperative frequency introduced by Arecchi and Courtens [4b]; Ω determines the characteristic scale for the SR field and time. As is seen, the effect of cavity losses consists of, first, eliminating that part of the Hamiltonian SR pulse (shown in Fig.1a) which resides at times $t > T_c$, and second, decreasing the intensity of the remaining part. It should be noticed that the latter retains patterns of the Hamiltonian pulse.

* Corresponding author: igoryzhov@yandex.ru

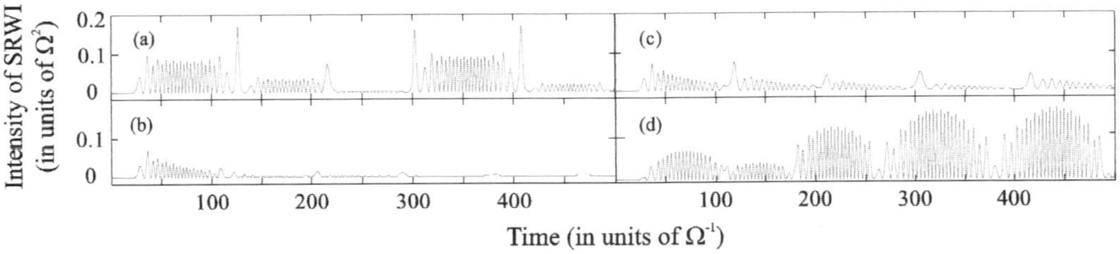

Fig. 1. (a) – The Hamiltonian SRWI pulse; (b) – $T_c = 50\Omega^{-1}$, $T_2 = T_2^* = \infty$; (c) - $T_2 = 50\Omega^{-1}$, $T_2 = T_2^* = \infty$; (d) - $T_2^* = 1000\Omega^{-1}$. $T_c = T_2 = \infty$. The initial population of the upper level $\alpha = 0.3$, расщепление дублета $\omega_{21} = 0.05\Omega$.

2. Homogeneous broadening. This type of broadening is characterized by the time T_2 and results in the relaxation of coherence of an individual emitter. In an open system, this leads to decreasing the SR signal along with shortening T_2. In the case of a cavity-assisted SRWI, the global effect of T_2 also consists of inhibition of SRWI for time moments $t > T_2$. However, it appears that even for $t > T_2$, there exists a visible signal (see Fig. 1b). We associate this effect with two facts: (i) the SR field although decreases due to the emitter's dephasing, but does not escape the cavity because of a high quality of the latter, (ii) the low frequency coherence does not decay in our model and provides an additional source for the high-frequency coherence. In combination, these two factors result in a partial revival of the SRWI signal for $t > T_2$.

3. Inhomogeneous broadening. This type of broadening differs from its homogeneous counterpart: the phase of each emitter is preserved and only the collective response decays with time. The phases of the emitters can in principle be inverted by applying a coherent pulse. If, however, the Q factor of the cavity is high enough. i.e. the SR field retains in the cavity for a long time, quasi-periodically/chaotically repeating combs of the SR pulses (like in our case) play the role of coherent trains, inverting the emitters' phases. Figure 1d shows the results of calculations of the SRWI dynamics for different values of the inhomogeneous relaxation time T_2^*. As is seen, even a significantly large (on the computation scale) time T_2^* leads to a noticeable change in the shape of the SRWI pulse as compared with an ideal (Hamiltonian) system (Fig. 1a). We attribute such a strong effect of T_2^* on the SR pulse to the fact that the system resides at the state of dynamic chaos already in the ideal (Hamiltonian) case. Therefore, even small variations in the emitters' phases during emission lead to an essentially different scenario of the system's optical dynamics, preserving, however, its chaotic nature. For shorter T_2^*, the dynamics of SRWI experiences even stronger changes.

Crystals doped with rare-earth ions (e.g., $LaF_3:Pt^{3+}$, $Y_2SiO_5:Pr^{3+}$, $Y_2SiO_5:Eu^{3+}$, and others) are those real objects in which the conditions for the observation of the nonstandard regimes of the cavity-assisted SRWI can be realized [4b].

This work was supported by the Russian Foundation for Basic Research (grant #15-02-08369).

References

[1] R.H. Dicke, Phys. Rev. **93**, 99 (1954)

[2] A.I. Zaitsev, V.A. Malyshev, E.D. Trifonov, and I.V. Ryzhov, JETP **88**, 278 (1999)

[3] I.V. Ryzhov, A.I. Zaitsev, and E.V. Shuval-Sergeeva, Opt. Spectrosc. **112**, 604 (2012)

[4] I.V. Ryzhov, N.A. Vasil'ev, I.S. Kosova, M.D. Shtager, and V.A. Malyshev, (a) Opt. Spectrosc. **120**, 440 (2016); (b) JETP **124**, 683 (2017)

EPJ Web of Conferences **161**, 02022 (2017) DOI: 10.1051/epjconf/201716102022
PECS-2017

The features of modelling semiconductor lasers with a wide contact

Alexey Rzhanov*

Lomonosov Moscow State University, Physics Department, Leninskie Gory, Moscow, 119991 Russia

Abstract. The aspects of calculating the dynamics and statics of powerful semiconductor laser diodes radiation are investigated. It takes into account the main physical mechanisms influencing power, spectral composition, far and near field of laser radiation. It outlines a dynamic distributed model of a semiconductor laser with a wide contact and possible algorithms for its implementation.

Semiconductor laser diodes (LD) appeared in the early 60-ies of the last century [1-2]. In 1969, due to double heterostructure implementation continuous wave (CW) LD operation at room temperature were obtained [3-4], and then the process of the LD technology development started. In the 80-ies of the last century the first works on mathematical modeling of LD dynamic and static behavior appeared [5-6].

One of the modern designs of the powerful LD with quantum well (QW) active layer and wide contact is shown in Fig. 1. Contact width w in the 8-10 W LD is about 100 μm.

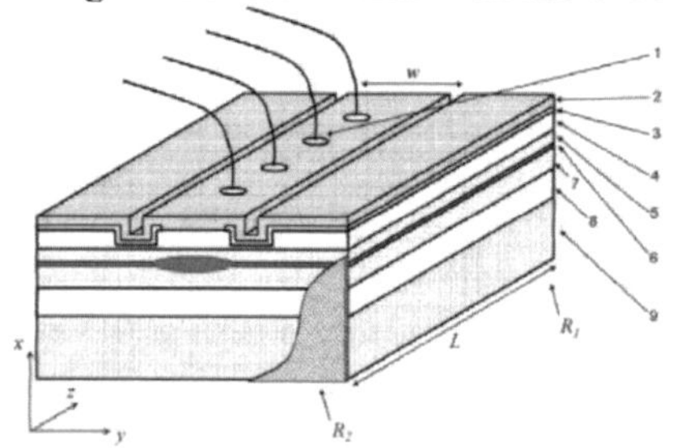

Fig. 1. Powerful injection laser: contacts (1), contact layer (2), insulating layer (3), P - emitter (4), waveguide layers (5,7), QW layer (6), N - emitter (8), substrate (9), mirrors (R1, R2) [7].

The basis of the LD models are kinetic (speed) equations that describe the balance of non-equilibrium injected carriers and photons in the active region of the laser. In their simplest form these equations write for the average values, not taking into account the spatial distribution of carriers and the optical field. In more detailed, self-consistent, models of LD the spatial interaction between the laser radiation and carriers takes into account. At other side, the waveguide permittivity profile and effective index depends on the concentration of carriers in active layer.

The number of kinetic equations in the model must match the number of lateral modes that have fallen through the generation threshold. The set of equations for carriers and

* Corresponding author: rjanov@mail.ru

photons concentrations along with a set of relations between the coefficients, form of the kinetic part of the model.

Second, optical part of the self-consistent model is the wave equation converted into a one-dimensional Helmholtz equation with complex variables:

$$\frac{d^2\psi_j(y)}{dy^2}+\left(\frac{\omega_j^2}{c^2}\varepsilon(y)-\beta^2\right)\psi_j(y)=0, \qquad (1)$$

where ω - optical frequency, β - longitudinal propagation constant, ψ - amplitude profiles of modes, ε(y) - effective permittivity, depending on the carrier concentration, c - the speed of light in vacuum.

The modern LD with output power of 10-20 W have the broad pumped region ("strip") - about w=100-200 μm width (see Fig. 1) and the spectral band is 4-5 nm. This is the main difficulty in creating an adequate model of LD with a broad contact.

In addition to the calculation of the transverse modes of one generation channel, for which a single set of kinetic equations and one wave equation are sufficient, in a broad cavity LD, you should consider the filamentation of radiation - decay of the optical wave at independent channels with unique light frequency and phase. As a result it is necessary to solve several equations (1) and the second kinetic equations for different channels with unknown in advance the size and frequency of the radiation. The reason for this phenomenon is the low coherence length of LD radiation (about 1-5 sm) which prohibits the phase locking between channels. Calculations and experiment show that at a cavity length of 1000 microns, and the width of the active region of the LD resonator of 100 μm not less than 3 channels is shaped [8-9]. The solution of this problem, associated with forced separation of the channels, does not lead to adequate results of calculations, although for some cases allows to qualitatively describing the dynamic pattern of switching between the channels and transverse modes [9]. The model as a self-consistent set of equations grows several times in this approach, as the number of equations (1) in the model must match the number of channels. The main question is determination of spatial dimensions of each channel, as well as the energy exchange between non phase-locked areas of generation. This task until now has not yet resolved.

A separate topic for consideration is the accounting processes associated with heating of the active region of LD. This requires the solution of the heat equation in parallel with the dynamic task. Such calculations in the form of full self-consistent problem yet due to the fact that it requires huge computer capacities. Thus, the separate problems of dynamics and thermal conductivity gives quite adequate results.

There are several physical mechanisms that influence the dynamics of LD radiation, for example, "spectral hole burning", the polarization effects due to light and heavy holes in QW stressed layers, and so on. However, to include in the model many mechanisms does not work due to the occurrence of instabilities and because of computer restrictions on the amount of computation.

References

[1] Z.I. Alferov, R.F. Kazarinov. Copyright № 181737. Patent № 950840 30.03.1963.

[2] H. Kroemer, Proc. IEEE **51**, 12 (1963)

[3] Z.I. Alferov et al., Sov. Phys. Tech. Semicond. **3**, 1328 (1969)

[4] I. Hayashi et al., Appl. Phys. Lett. **17**, 2 (1970)

[5] J. Buus, IEEE J. Quant. Electron. **QE-18**, 7 (1982)

[6] J. Buus, IEEE J. Quant. Electron. **QE-19**, 6 (1983)

[7] I.S. Tarasov, Quant. Electron. **40**, 8 (2010)

[8] O.I. Koval, A.G. Rzhanov, G.A. Solovyev, Phys. of WP, **21**(4), 287 (2013)

[9] V.V. Bliznyuk, M.A. Brit, I.S. Gadaev et al., Bull. RAS: Physics **79**, 12 (2015)

Exact dynamics of a two-level atom beyond the rotating wave approximation

Vitalii Semin*

Samara National Research University, Department of nanoengeeniring, 34 Moskovskoe Shosse, Samara, 443086 Russia

Abstract. Interaction Hamiltonians of some models beyond the rotating wave approximation are just a product of two commuting operators. The evolution operator of such models can be transformed into product of two independent chronological exponents with the help of Hubbard-Stratonovich transformation. We use such a representation of the evolution operator to exactly describe a two-level atom in a photonic thermostat.

Description of dynamics of an open quantum system is an extremely difficult task. Today there are only a few models allowing one to find the exact dynamics. Among these are a quantum harmonic oscillator in a thermostat of oscillators [1], Garraway's model [2], pure dephasing of a two-level atom [3], a fermionic oscillator in a bath of fermionic oscillators [4], spin star models [5]. The characteristics of the abovementioned models that allow us to find the exact dynamics are some symmetries of the model Hamiltonian or a possibility to truncate the Hilbert space of the model.

In this paper we consider the model of a free relaxation of a two-level atom beyond the rotating wave approximation. The interaction Hamiltonian in the interaction pictures writes as a product of two independent operators, namely

$$H_I(t) = \left(\sigma_- e^{-i\omega_0 t} + \sigma_+ e^{i\omega_0 t}\right)\left(\sum_k g_k \left(b_k e^{-i\omega_k t} + b_k^+ e^{i\omega_k t}\right)\right) = A(t)B(t), \qquad (1)$$

where $\sigma_\mp$ is the Pauli matrix, ω_0 is the atomic transition frequensy, $b_k(b_k^+)$ is annihilation (creation) operator of kth photon in the bath with frequency ω_k and g_k is the interaction constant.

The exact dynamics is described by the reduced density operator

$$\rho_A = Tr_B\left(U(t)\rho(0)U^+(t)\right), \qquad (2)$$

where the trace is taken over the bath and the evaolution operator satisfies the Schroedinger equation with the Hamiltonian (1). We also assume that initially the bath and the atoms are uncorrelated and the bath is in the thermal equilibrium state.

Now we can use the fact of commuting operators in (1). For such operators the evolution operators in (2) can be represented as a product of two independent chronological exponents with the help of the Hubbard-Stratonovich transformation. After the transformation the trace in (2) can be calculated explicitly and the result expression has the form

$$\rho_S(t) = \int \mathcal{D}\left(z_1(t), z_2(t)\right)\overleftarrow{T}\exp\left[-i\int_0^t z_1(s)A(s)ds\right]\rho_A(0)\overrightarrow{T}\exp\left[i\int_0^t z_1(s)A(s)ds\right], \qquad (3)$$

* Corresponding author: semin@ssau.ru

where the arrows indicate the time-ordering and $\mathcal{D}$ is some functional measure. The expression (3) is exact and its calculation is the main result of this research.

References

[1] B.L. Hu, J. Paz, Y. Zhang, Phys. Rev. D **45**, 1576 (1992)
[2] B.M. Garraway, Phys. Rev. A **55**, 2290 (1997)
[3] W.G. Unruh, Phys. Rev. A **51**, 992 (1995)
[4] A. Ghosh, S.S. Shinha, D.S. Ray, Phys. Rev. E **86**, 011138 (2012)
[5] H.P. Breuer, D. Burgarth, F. Petruccione, Phys. Rev. B **70**, 045323 (2004)

EPJ Web of Conferences **161**, 02024 (2017) DOI: 10.1051/epjconf/201716102024
PECS-2017

The tunnelling of electromagnetic radiation in chirped dielectric structure

Anatoliy Kozar, *Vladimir* Marchenko, and *Pavel* Shestakov*

Lomonosov Moscow State University, Physics Department, Leninskie Gory, Moscow, 119991 Russia

Abstract. The tunnelling of optical beams through chirped Bragg grating is considered. Lateral shifts of transmitted and reflected beams were calculated. The saturation of Goos-Hanchen shift of transmitted light beams is shown.

The chirped Bragg grating (CBG) is the structure with linear deviation of its period along propagation direction:

$d(z)=d_0+\alpha z$ (d_0 – grating period, α – deviation parameter), (1).

The use of CBG for dispersion compensation and amplification of chirped pulses is based on its reflective properties [2]. When we deal with spatial beams a CBG allows obtaining after reflection either divergent or convergent light beams. The tunnelling of optical radiation decreases an efficiency of a CBG since it leads to power loss in reflection. Along with that the transmission of electromagnetic radiation through such structures allow to model quantum effects, which direct experimental observation is not available so far (3, 4).

It is well known that CBGs have local forbidden gaps not only in frequency and spatial spectra but also along the longitudinal coordinate – so called photonic barrier (PB). The PB boundaries are determined from the expression $\delta = \pm\kappa$, where the detuning from the Bragg resonance $\delta = k_0\varepsilon_0^{1/2}\cos\theta_0-\pi/d_0+\beta(z-z_0)$, $\kappa = \Delta\varepsilon/4k_0\varepsilon_0^{1/2}$ – the Bragg coupling, $k_0 = \pi/d_0$, $\beta = -\alpha\pi/(2n_0d_0^2)$, $\varepsilon_0 = n_0^2$, $\Delta\varepsilon \approx (\varepsilon_1 - \varepsilon_2)/2$, $\Delta\varepsilon/\varepsilon_0 << 1$ - the dielectric contrast, n_0 – the average refractive index. The length of the PB is determined by the formula $l_{PB} = 2\kappa/|\beta|$.

Fig. 1 shows the oblique incidence ($\theta_0=30^o$) of a weakly diverging beam with gaussian profile under the conditions when the photonic barrier with $l_{PB} = 8\ \mu m$ is located in the $L/2$ vicinity. There is virtually no profile distortion and phase modulation of the light beam. The total lateral shift (segment CD = $35\ \mu m$) of the reflected beam at the boundary $z = 0$ is composed of a shift due to the beam propagation in the transparency region of the chirped structure and a shift at the photonic barrier surface ($AB = 2\ \mu m$), or the Goos – Hänchen shift.

The oblique incidence of a light beam on a finite periodic structure without period deviation ($\alpha = 0$) is accompanied by the Goos-Hänchen shift of reflected and transmitted beams when the angular spectrum of the beam is within the band gap [5]. Since the phase of the reflection coefficient $\phi_R = \arctan(\delta/q \tanh \kappa L)$ the Goos-Hanchen shift is determined by $\Delta_R = d\phi_R/dk_x = \tanh(\kappa L)/\kappa \tan\theta$ when $\delta = 0$, where $q = (\delta^2 - \kappa^2)^{1/2}$. Assuming that the light reflects not from the surface of the periodic structure but from the effective depth ($d_{eff} = 1/2\kappa$) inside this structure we simplify the expression for the Goos-Hänchen shift as $\Delta = 2d_{eff} \tan\theta$.

These estimations are valid for semi-infinite structures, i.e. the attenuation inside the PB should be sufficient for the reflection coefficient is about unity. Because of this the length of

* Corresponding author: iveage@physics.msu.ru

the PB is restricted by the condition $tan(\kappa l_{PB}) \approx 1$, which means that $\kappa l_{PB} \geq 3$. When a light beam is incident onto a structure with $\kappa l_{PB} < 3$, a noticeable tunnelling of electromagnetic radiation through the PB is observed. This regime reduces the efficiency of the compression–stretching mechanism of reflected pulses (beams).

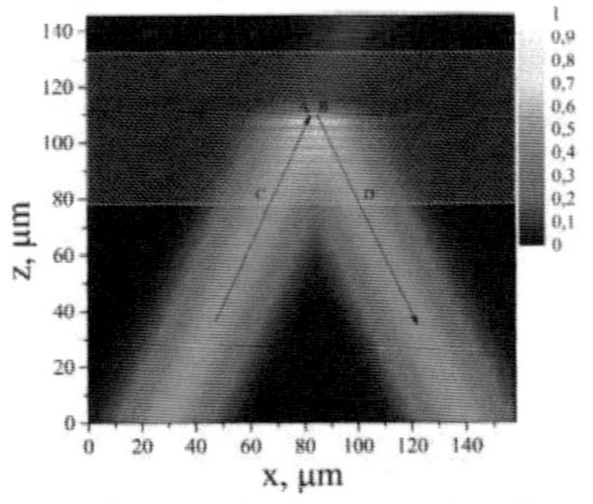

Fig. 1. Reflection of a wide light beam from a photonic barrier ($l_{PB} = 8\ \mu m$) located inside the periodic structure ($\alpha = 2\cdot10^{-3}$, $n_1 = 1.6$, $n_2 = 1.4$, $\theta_0=30^o$). The arrows indicate the direction of beam propagation.

The value of the Goos-Hänchen shift of a transmitted beam is estimated numerically by the expression:

$$\Delta_T=\int x|E_{tr}|dx \,/ \int|E_{tr}|dx \qquad (1)$$

where E_{tr} – the amplitude of forward propagating wave, x – the traverse coordinate. The Fig. 2 shows the calculated values of the Goos-Hänchen shift in dependence on periodic structure for the different values of photonic barrier lengths located in the $L/2$ vicinity.

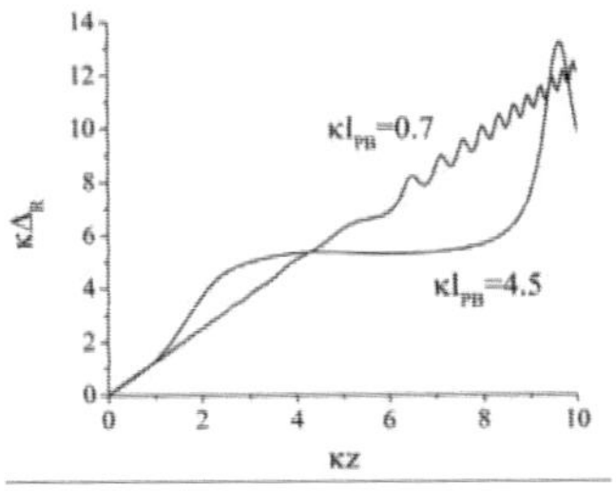

Fig. 2. The values of the Goos-Hänhen shift of a transmitted beam in dependence on CBG length ($\alpha = 2\cdot10^{-7}$, $L=2\ cm$, $n_0 = 1.4$, $\Delta\varepsilon/\varepsilon_0=10^{-3}$).

When $\kappa l_{PB} < 3$ the shift of a transmitted beam increases linearly with length of the structure. If the length of the photonic barrier is quite large ($\kappa l_{PB} > 3$) the values of the Goos-Hänchen shift saturates. This effect is analogous to the time delay saturation of tunnelled pulses through a periodic structure without a period deviation ($\alpha = 0$). It was noted that the saturation of time delay happens because of energy storage in a near-surface layer of the periodic structure [6].

This study is supported by the Russian Science Foundation (Grant No 17-11-01157).

References

[1] S. Kaim, S. Mokhov, B. Y. Zeldovich, L.B. Glebov, Opt. Eng. SPIE **5**, 53 (2014)
[2] R. Sun, D. Jin, F. Tan et al., Opt. Express **24**, 20 (2016)
[3] S. Longhi, Phys. Res. Int. **2010**, 645106 (2010)
[4] P.B. Wilkinson, Phys. Rev. E **65**, 5 (2002)
[5] D. Felbaq, A. Moreau, R. Smaâli, Opt. Lett. **28**, 18 (2003)
[6] H. Winful, Opt. Express **10**, 25 (2002)

Nonlinear properties of high-Q optical microresonators in normal dispersion range

Artem Shitikov[1,2,*], *Valery* Lobanov[1], *Nikolay* Pavlov[1,3], *Andrey* Voloshin[1], *Igor* Bilenko[1,2], and *Michael* Gorodetsky[1,2]

[1]Russian Quantum Center, Moscow, Russia
[2]Lomonosov Moscow State University, Faculty of Physics, Moscow, Russia
[3]Moscow Institute of Physics and Technology, Dolgoprudny, Russia

Abstract. We demonstrate the generation of Kerr frequency combs and platicons in whispering gallery mode crystalline microresonators in normal group velocity regime at 780 nm and 1064 nm wavelengths.

Whispering gallery mode (WGM) microresonators can have extremely high values of quality-factor (exceeding 10^9) in the optical range, that allows low threshold of many interesting nonlinear effects, in particular, the generation of Kerr frequency combs [1]. Kerr combs in optical microresonators may become a basis for a new class of compact devices for spectroscopy, precision measurements, and generation of femtosecond pulses.

Coherent low-noise frequency combs are of particular interest. These combs may be generated in the form of dissipative Kerr solitons in the region of anomalous group velocity dispersion (GVD) [2] and to date have been already demonstrated on different platforms. However, most materials suitable for the manufacturing of WGM microresonators (for example, widely used fluorides MgF_2 and CaF_2) have a normal GVD in the visible and near IR ranges, which prevents soft generation of coherent frequency combs and bright dissipative Kerr solitons. Therefore, the development of new methods for coherent frequency comb excitation in microresonators with normal GVD and studying the properties of such combs is an actual and practically significant task. It was shown recently that in the normal GVD regime coherent frequency combs could be achieved in the form of soliton-like pulses of a particular form, so-called platicons [3]. It was also demonstrated that the efficiency of the transformation of pump power into the power of the frequency comb lines is significantly better for platicons compared to the generation of bright solitons in anomalous GVD regime. This properties of platicons are important for many practical applications.

In the recent years, several methods for the generation of platicons were proposed: the use of a local dispersion perturbations caused by mode interactions [3,4], pump modulation at a frequency equal to the free spectral range (FSR) of a microresonator and biharmonic pump [5], self-injection locking of a laser source to a mode of the WGM microresonator [6].

In our work MgF_2 microresonators and prism coupling were used. For 3 mm in diameter MgF_2 microresonators the GVD is mostly determined by material dispersion which is normal at given wavelengths. The microresonator's loaded quality-factor exceeded $5x10^8$.

For the pump at 1064 nm, the experimental setup was adapted both for biharmonic and for the amplitude-modulated pump (see Fig. 1). For the pump at 780 nm, a setup with an

[*] Corresponding author: Shartev@gmail.com

external cavity diode laser self-injection-locked to a mode of the same high-Q WGM microcavity was used (Fig. 1, on right). The experimental results showed that, due to the effect of self-injection locking, the generated linewidth may become less than 1 kHz.

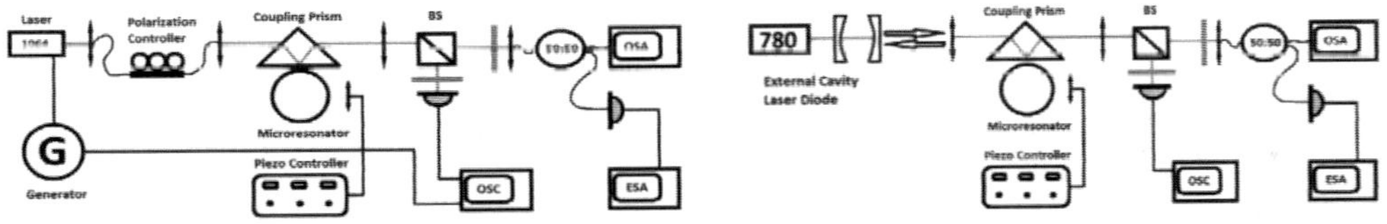

Fig. 1. Experimental setup for the pump at 1064 nm; self-injection locked pump at 780 nm (right).

For the pump at 1064 nm, combs were observed for all modes with a quality-factor exceeding $5x10^8$ and coupling efficiency better than 10%. Frequency combs with asymmetric spectral envelope caused by mode interaction were observed. High-efficiency power conversion into comb lines (more than 30%) was also observed.

For the pump at 780 nm, in addition to the combs similar to the combs obtained at 1064 nm, platicon-like spectra with a width of 1-3 nm were also regularly detected. The characteristic plateau of the spectrum was 20 dBm below the pump line. The width of the microwave beatnote signal between the comb lines was less than 1 kHz and was detected on an electrical spectrum analyzer at a frequency corresponding to the FSR of microresonator.

The results of the experiments are shown in Fig. 2.

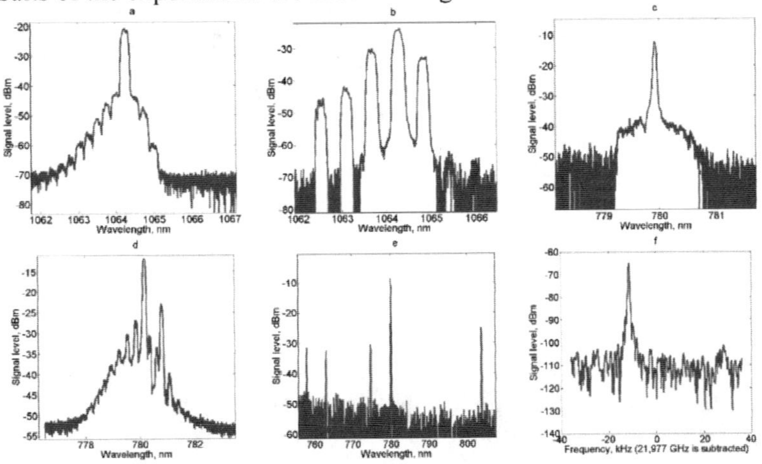

Fig. 2. a, b – frequency combs, formed as a result of mode interaction with pump at 1064 nm: a – the distance between the lines corresponds to 3 FSR; b – the first lines are below the pump by 6 dB, the generation efficiency is more than 30%. c, d, e – combs generated by the self-injection-locked diode laser pumping at 780 nm. c – is a platicon-like spectrum with a width of 1.5 nm. d – platicon-like spectrum with pronounced comb lines. e – hyperparametric generation in the range of 50 nm. f – the beatnote between the comb lines, the line width is less than 1 kHz.

Experimental setups for the of optical frequency combs generation in WGM microresonators with normal GVD were designed and tested. It is planned to investigate the combs properties and to compare different methods of its excitation.

This work was supported by the Russian Science Foundation (project №17-12-01413).

References

[1] T. Herr, K. Hartinger et al., Nature Photon. **6**, 48 (2012)
[2] V.E. Lobanov, G. Lihachev et al., Opt. Exp. **24**(24), 27382 (2016)
[3] V.E. Lobanov, G. Lihachev et al., Opt. Exp. **23**(6) 7713 (2015)
[4] X.X. Xue, Y. Xuan et al., Laser Photon. Rev. **9,** L23 (2015)
[5] V.E. Lobanov, G. Lihachev, M. Gorodetsky, Europhys. Lett. **112**, 5 (2015)
[6] W. Liang, A.A. Savchenkov et al., Opt. Lett. **39**, 2920 (2014)

Preparation of single-photon states via cavity-assisted spontaneous parametric down-conversion for quantum memory based on $Y^7LiF_4:Nd^{3+}$ crystal

Dmitrii Akatiev, *Ilnur* Latypov, *Andrey* Shkalikov*, and *Alexey* Kalachev

Kazan E. K. Zavoisky Physical-Technical Institute, 10/7 Sibirsky tract Str., Kazan, 420029 Russia

Abstract. We report on the realization of a tunable single-photon source compatible with quantum memories based on isotopically pure Y^7LiF_4 crystals doped with Nd^{3+} ions. The source is based on spontaneous parametric down-conversion in a PPLN crystal placed in a resonator. The latter is formed by two mirrors which are transparent for the pump (532 nm) and idler (1377 nm) fields but high reflective for the signal field (867 nm). The width of the second-order cross-correlation function between the signal and idler photons is determined to be 1.5 ns for the cavity length of 8 cm, which corresponds to a cavity bandwidth of 100 MHz.

1 Introduction

Single-photon sources are believed to be an important ingredient for long-distance quantum communication and other applications of quantum information science. For getting through the fibre losses and implementing long-distance quantum communication protocols, one can take advantage of quantum repeaters [1, 2], which in turn requires storage of nonclassical light states in a quantum memory device. Among different schemes of quantum memory (see, e.g., recent reviews [3, 4]), off-resonant solid-state Raman schemes might be useful. In this respect, isotopically pure crystal $Y^7LiF_4:Nd^{3+}$ seems to be one of the promising materials [5, 6]. The present work is devoted to developing a heralded single-photon source that is compatible with such a quantum memory device. In doing so, we follow the approach of [7] and make the spectral width of generated single-photon states comparable with the quantum memory bandwidth (10–100 MHz) by using a single-resonant optical parametric oscillator (OPO) operating far below threshold.

2 Basic results

For the parametric interaction in our OPO resonator we use a 20 mm long PPLN crystal (Labfer) that supports quasi-phase matching for type-0 SPDC to convert photons at 532 nm

* Corresponding author: andrei_vs@rambler.ru

to photon pairs, with one photon located at 867 nm (signal field) and the other at 1377 nm (idler field). The former corresponds to the resonant transition of Nd^{3+} ions in the Y^7LiF_4 crystal. Temperature of the nonlinear crystal is controlled to the precision of 0.01°C. The mirrors of the resonator are high reflective only for the signal field (R ≈ 99.8% and 97%), which corresponds to the signal-resonant OPO. The sides of the PPLN are antireflection coated to minimize losses within the cavity. In order to perform cross-correlation measurements between signal and idler photons, each of them is guided to an avalanche photodetectors connected to a coincidence circuit. Filtering of the pump field by several orders of magnitude in each channel is accomplished by a sequence of a diffraction grating, a cut-off dichroic mirror, an interference filter, and also a Bragg fibre filter. An example of the measured second-order cross-correlation function between signal and idler photons is shown in Fig. 1. In this case, the cavity length is equal to 8 cm so that the correlation function broadens from 300 ps, which is observed for SPDC in the free space, to 1.5 ns, which corresponds to the spectral linewidth of 100 MHz. The modulation of the decay curve reflects contribution of adjacent cavity modes and corresponds to the cavity round-trip time of 0.8 ns.

The work is supported by Russian Science Foundation (Grant No. 14-12-00806).

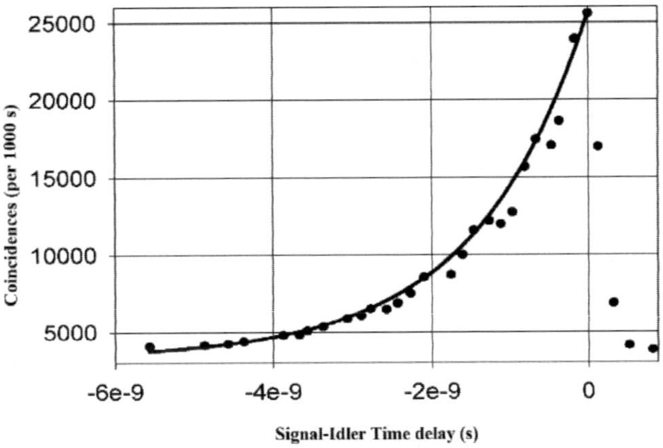

Fig. 1. Second-order cross-correlation function between signal and idler photons.

References

[1] H-J. Briegel, W. Dür, J.I. Cirac, P. Zoller. Phys. Rev. Lett. **81**, 5932 (1998)
[2] N. Sangouard, C. Simon, H. de Riedmatten, N. Gisin. Rev. Mod. Phys. **83** 33 (2011)
[3] F. Bussières et al., Journal of Modern Optics **60**, 1519 (2013)
[4] K. Heshami et al., Journal of Modern Optics **63**, S42 (2016)
[5] R.A. Akhmedzhanov et al., Las. Phys. Lett. **13,** 015202 (2016)
[6] R.A. Akhmedzhanov et al., Las. Phys. Lett.**13,** 115203 (2016)
[7] M. Scholz, F. Wolfgramm, U. Herzog, O.Benson. Appl. Phys. Lett. **91** 191104 (2007)

EPJ Web of Conferences **161**, 02027 (2017)
PECS-2017
DOI: 10.1051/epjconf/201716102027

Light bullets formation accompanied by odd harmonics generation under Mid-IR femtosecond filamentation in LiF

Evgeniya Smetanina[1,*]

[1] *CELIA (Centre Lasers Intenses et Applications) Université Bordeaux Domaine du Haut-Carré 43 rue Pierre Noailles 33405 Talence Cedex France*

Abstract. The formation of light bullets in mid-IR femtosecond filamentation is investigated numerically by means of the unidirectional pulse propagation equation (UPPE). A general approach for light bullets formation is presented for condensed media under anomalous GVD. The main parameters of the light bullets are evaluated. The features of super continuum and odd harmonic generation accompanying the light bullets formation are analyzed.

An appearance of new hight-power femtosecond laser sources in MID-IR wavelength region [1] leads to an active research of MID-IR laser filamentation regimes in both, condensed transparent dielectrics [2, 3] and gases [4–6]. The wavelengths from 2.5 to 4 μm correspond to the anomalous group velocity (GVD) region in a large number if optical materials (silica glass, CaF_2, BaF_2, LiF, YAG). Thus, an interplay between Kerr nonlinearity and anomalous GVD leads to a formation of spatially and temporally compressed wavepackets - light bullets (LB) [7–9]. In the current work we present the numerical results on the filamentation of 3.1 μm femtosecond laser pulse in LiF. The formation of the LB and evolution of the pulse spectrum will be considered.

The modeling was performed by means of unidirectional pulse propagation equation. Two $\chi^{(3)}$ – *based* effects are included: Kerr nonlinearity and third harmonic generation, that leads to a cascade generation of odd harmonics. We consider a weakly focused linearly polarized 3.1-μm laser pulses with peak power P=1.5 P_{cr}, where the critical power of the self-focusing is P_{cr} equals to 100 MW.

We report, that the main spatio-temporal features of the LB formation at 3.1μm in LiF are similar to those for LBs formation at 1.8μm in fused silica [8]. The duration of the LB reaches the value about 2-3 optical cycles (Fig.1 a-b). The odd harmonic generation accompanies the pulse propagation and the LB formation. The peak intensity of the harmonics oscillates along the propagation distance with the period corresponding to the phase matching conditions. The obtained results are close to the odd harmonic generation during filamentation in fused silica and CaF_2 [3]. The closer the pulse is to the nonlinear focus, the wider are the harmonics spectra. Finally, the harmonics spectra are overlapped with the widely broadened blue wing of supercontinuum that appears due to the self-steepening of the pulse and corresponds to the LB formation (Fig.1 c). The beating of the harmonics intensity along the propagation distance provides the beating of the pulse peak intensity in spatio-temporal domain (Fig.1 d). Together with that, the generated odd harmonics may lead to appearance of new ionization and color-centers formation pathways with lower number of photons.

*e-mail: evgeniya.smetanina@u-bordeaux.fr

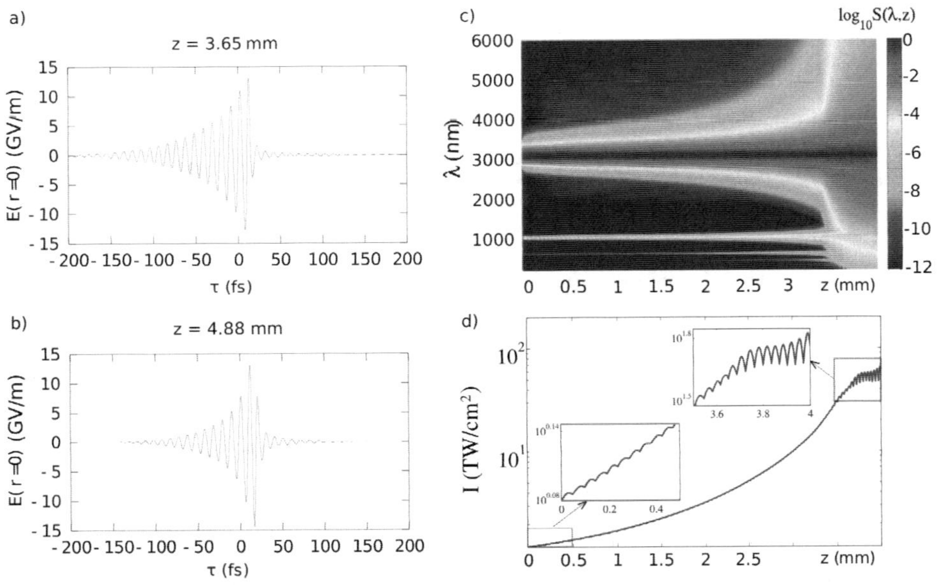

Figure 1. a)-b) LB time profiles of the electric field E(τ,r=0), the propagation distance z is specified on the figure, c) evolution of the angularly integrated spectrum $S(\lambda, z)$ along the laser pulse propagation axis z under filamentation in LiF. $\tau < 0$ corresponds to the pulse front. $S(\lambda, z)$ is normalized on its maximum value at each z. The incident pulse parameters are a_0=90μm, τ_0 = 120fs in a) and τ_0 = 60fs in b), the focal distance is 5mm.

Prof.V.P. Kandidov is acknowledged for fruitful discussions. Prof. S.Skupin is acknowledged for the opportunity to use the UPPE cod "CUPRAD". This work was granted access to the HPC resources of TGCC under the allocation A0010506129 made by GENCI.

References

[1] R. Butkus, R. Danielius, A. Dubietis, A. Piskarskas, A. Stabinis, Applied Physics B **79**, 693 (2004)

[2] S.V. Chekalin, V.O. Kompanets, A.V. Kuznetsov, A.E. Dormidonov, V.P. Kandidov, Laser Physics Letters **13**, 065401 (2016)

[3] N. Garejev, V. Jukna, G. Tamošauskas, M. Veličkė, R. Šuminas, A. Couairon, A. Dubietis, Opt. Express **24**, 17060 (2016)

[4] S.V. Chekalin, A.E. Dokukina, A.E. Dormidonov, V.O. Kompanets, E.O. Smetanina, V.P. Kandidov, Journal of Physics B: Atomic, Molecular and Optical Physics **48**, 094008 (2015)

[5] P. Panagiotopoulos, P. Whalen, M. Kolesik, J.V. Moloney, Nat Photon **9**, 543 (2015)

[6] N.A. Panov, D.E. Shipilo, V.A. Andreeva, O.G. Kosareva, A.M. Saletsky, H. Xu, P. Polynkin, Phys. Rev. A **94**, 041801 (2016)

[7] L. Bergé, S. Skupin, Phys. Rev. Lett. **100**, 113902 (2008)

[8] E.O. Smetanina, V.O. Kompanets, A.E. Dormidonov, S.V. Chekalin, V.P. Kandidov, Laser Physics Letters **10**, 105401 (2013)

[9] I. Gražulevičiūtė, G. Tamošauskas, V. Jukna, A. Couairon, D. Faccio, A. Dubietis, Opt. Express **22**, 30613 (2014)

Effects of manifestations of the initial quantum correlations in the coherent scattering of the atom in the standing wave

Andrey Trubilko*

St.-Petersburg University of the State Fire Service of the Ministry of Emergency Situations of Russia, 196105, St.-Petersburg, Russia

Abstract. Some effects of the initial quantum correlations between the atom and the quantum mode of the standing wave of electromagnetic field in the condition of coherent scattering are considered.

A number of quantum effects of the coherent scattering atom in the field of a standing low frequency electromagnetic wave [1] is demonstrated. Here, the nature of the quantum dynamics of a two-level atom strongly depends on the initial conditions of the preparation of interacted systems [2-4]. This initial separable and inseparable state of systems are prepared by broadband parametrical sources. The effective scattering optical potential is made by a quantized field and two coherently classical counter-advancing waves of the optical frequency range. The interaction of the atom with each field happens in the wing of the absorption line of the working transition. However, we assume that the conditions of the Raman and two-photon resonances at the same transition are satisfied. Under these approximations the optical quantum scattering potential can be presented in the form of

$$U(x) = g^2 h^{-1} [\Pi_+(\omega_c)(a^+aS_z + S_z) + \Pi_-(\omega_c)(2)^{-1}] \sin^2(kx). \quad (1)$$

Here g is coupling constant of the interaction between the quantum cavity mode and a two-level atom, the operator a^+a is number of photon operator for cavity mode with frequency ω_c and wave vector k is wave vector, S_z is the inversion operator for two level atom with transition frequency ω_0. The spectroscopy functions is $\Pi_\pm(\omega)=(\omega_0-\omega)^{-1}\pm(\omega_0+\omega)^{-1}$. Because the physical values in (1) are the integrals of motions, thus the average values are constant, and they are defined by initial conditions. We demonstrate that in the scattering of the atom only on quantized standing wave, scattering is absent, when the atom and the field are independent and thus the initial state is separable. On the contrary, the absence of scattering for the initial inseparable state is observed when the optical potential is formed by all three electromagnetic fields.

The effect of the quantum deviation of the atom is discovered. It is observed when the optical potential is produced by the quantum standing wave and the coherently classically field, localized in a small space. This space is produced by the optical fiber. The optical potential for this case is determined by conditions of a two-photon interaction. And in this case the different nature of the dynamics of the atom is captured by the optical potential,

* Corresponding author: trubilko.andrey@gmail.com

different ways of preparing the initial atomic field system is addressed. In particular, if the wave functions in the coordinate representation have the Gaussian packet, the average position of the interacting atom and the average values of variances of the coordinate and momentum are determined by the initial conditions.

References

[1] A.P. Kazantsev, G.I. Surdutovich, V.P. Yakovlev, *Mechanical Action of Light on Atoms* (Singapore, London, World Scientific, 1990)
[2] A.I. Trubilko, JETP Lett. **98**, 249 (2013)
[3] A.I. Trubilko, JETP **123**, 557 (2016)
[4] A.I. Trubilko, JETP Lett. **105**, 617 (2017)

Femtosecond filamentation of optical vortex in a medium with anomalous group velocity dispersion

Evgeny Vasilyev, *Svyatoslav* Shlenov*, and *Valery* Kandidov

Lomonosov Moscow State University, Faculty of Physics and International Laser Center, 119991 Moscow, Russia

Abstract. Within the framework of the slowly varying wave approximation method, the pictures of spatio-temporal dynamics of femtosecond pulse in a ring beam with phase singularity were obtained during propagation in CaF_2 crystal with anomalous group velocity dispersion. We show formation of tubular structure of the beam and explain reasons of its appearance. Fluence and linear plasma concentration dependencies on propagation distance are analyzed.

Filamentation of laser radiation is a phenomenon of formation of a long spatiotemporal structure with high power density [1]. Femtosecond filamentation has been widely studied for Gaussian and other beams with a smooth phase [2, 3]. Filamentation of circular beams with phase dislocations can be promising for such applications as tubular refractive index micromodifications, electrons accelerations, etc [4]. We have numerically investigated the filamentation of a femtosecond laser pulse in fused silica for the case of an annular beam with a phase singularity at a wavelength of 800 nm [5]. In this paper, we study propagation of a pulse in mid-IR in circular beam with phase dislocation in presence of anomalous group velocity dispersion typical for calcium fluorides.

A numerical simulation of the problem was performed based on the system of differential equations for the slowly varying complex envelope of laser field $A(r,t,z)$ and free-carrier concentration $N_e(r,t,z)$:

$$\begin{cases} 2ik_0\dfrac{\partial A}{\partial z} = \hat{T}^{-1}\left[\left(\dfrac{1}{r}\dfrac{\partial}{\partial r}\left(r\dfrac{\partial}{\partial r}\right) - \dfrac{m^2}{r^2}\right)A\right] + \widehat{D}[A] + \dfrac{2k_0^2}{n_0}(\Delta n_k - \Delta n_{pl})A - ik_0(\sigma N_e + \alpha + \delta)A, \\ \dfrac{\partial N_e}{\partial \tau} = R_E(N_0 - N_e) + N_e(\nu_i - \beta), \end{cases} \quad (1)$$

where $\hat{T}$ is operator of wave nonstationarity [1], describing self-steepening of wave front, $\widehat{D}$ - dispersion operator, which is calculated in spectral domain according to Sellmeyer's formula. The model takes into account beam diffraction and pulse dispersion, Kerr's and plasma nonlinearities, Bremsstrahlung effect, nonlinear absorption and extinction, avalanche ionization and electron recombination.

The initial condition represents a circular beam with located in the center phase dislocation in transform-limited Gaussian pulse.

* Corresponding author: s.shlenov@phys.msu.ru

$$A(r,t,0) = A_0 \left(\frac{r}{r_0}\right)^2 e^{-r^2/2r_0} e^{-t^2/2t_0} e^{im\varphi},$$

where $r_0 = 230\ \mu\text{m}$ – effective beam radius, $t_0 = 60$ fs – pulse duration, $m = 2$ – topological charge, central wavelength is $\lambda_0 = 3000$ nm. The peak pulse power exceeds optical vortex self-focusing critical power P_{cr} [6] six times.

Numerical calculations show that the pulse propagation with specified parameters starts with its self-focusing, remaining annular structure of the beam (Fig.1). The peak intensity significantly increases, the laser pulse front steepens and the beam annular structure becomes thinner. Once the intensity reaches a value of about 5×10^{13} W/cm^2 at a distance of 1.7 cm from the beginning of the nonlinear medium strong ionization yields plasma appearance and, consequently, beam defocusing. At this distance linear plasma concentration reaches $\rho \sim 4 \times 10^{13}$ cm^{-1}, fluence being more than 0.12 J/cm^2. Beam defocusing leads to the ring structure becoming wider and peak intensity decreasing. Part of the optical energy flows to the beam periphery, the other one still forming a tubular structure.

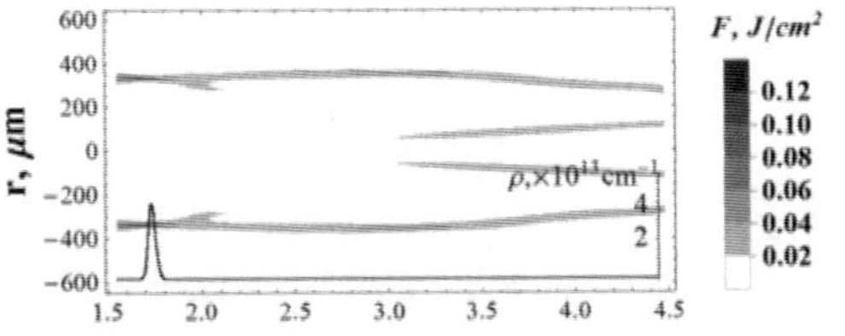

Fig. 1. Spatial distribution of fluence F and linear plasma concentration ρ. Distance z calculates from the leading edge of crystal.

Anomalous group velocity dispersion (AGVD) at the beginning makes pulse to compress and move with group velocity at the carrier frequency. Pulse envelope maintains a Gaussian-like shape, while the beam has an annular structure. The formation of such cylindrical spatiotemporal structure is reinforced by the influence of AGVD, because it prevents pulse spreading making high frequencies propagate with group velocity greater than group velocity of lower frequencies in the pulse. Self-focused intensive annular part of the beam propagates in the medium with higher refractive index. This leads to the fact that it is shifted backwards in time relative to the initial center of the pulse. Simultaneously optical energy flows towards optical axis due to linear diffraction. The presence of a phase singularity in the beam prevents energy transformation directly to optical axis and, as a result, the second annular structure of smaller diameter appears at distance 3.2 cm. At the same time, radiation in outer ring refocuses. Tubular structure of the beam remains at least up to 4.5 cm, wherein outer and inner rings move towards each other.

In summary, we have presented the results of numerical calculations of optical vortex femtosecond filamentation in mid-IR based on mathematical model with slowly varying wave approximation. We have obtained and analyzed the spatio-temporal dynamics of pulse propagation and formation of tubular structure of the beam with pulse time compression due to anomalous group velocity dispersion.

This work has been supported by the Council of RF President, grant NSh 9695.2016.2.

References

[1] V.P. Kandidov, S.A. Shlenov, O.G. Kosareva, Quant. Electron. **39**, 3 (2009)
[2] A. Couairon, A. Lotti, P. Panagiotopoulos et al., Proc. of SPIE **8770**, 87701E (2013)
[3] S.V. Chekalin et. al., J. Phys. B: At. Mol. Opt. Phys. **48**, 9 (2015)
[4] P. Sprangle, E. Esarev, J. Krall, Physics of Plasma **3**, 2183 (1996)
[5] E.V. Vasil'ev, S.A. Shlenov, Quant. Electron. **46**, 1002 (2016)
[6] L.T. Vuong, T.D. Grow, A. Ishaaya, A.L. Gaeta, Phys. Rev. Lett. **96**, 13 (2006)

EPJ Web of Conferences **161**, 03001 (2017) DOI: 10.1051/epjconf/201716103001
PECS-2017

Combined analysis of the temperature dependences of the fluorescence spectra and images of single molecules in polymer films

Tatiana Anikushina[1,2,*], *Andrei* Naumov[1,2], and *Lothar* Kador[3]

[1]Moscow State Pedagogical University, 29/7 Malaya Pirigivskaya Str., Moscow, 119991 Russia
[2]Institute for Spectroscopy RAS, 5 Fizicheskaya Str., Troitsk, Moscow, 108840 Russia
[3]Bayreuth University, 30 Universitätsstraße, Bayreuth, 95447 Germany

Abstract. In this poster talk we discuss a possibility to realize the combined analysis of the temperature dependences of the zero-phonon line widths and fluorescence images of single molecules in thin polymer films. Such analysis is aimed to obtain the local characteristics of low-temperature vibrational dynamics (individual parameters of quasi-localised low-frequency vibrational modes) in relation to the sample structure.

One of the most topical interdisciplinary scientific fields has recently become the single molecule spectromicroscopy (SMSM) in condensed matter [1] This method opens the possibility of investigating the internal dynamics of the medium at the microscopic level and allows the diagnostics of the materials with an ultra-high spatial resolution [2, 3].

This method acquires special capabilities in the detection of electron-vibrational spectra of impurity molecules, embedded as nanoprobes into a solid medium, at low (cryogenic) temperatures [1-4]. At such temperatures (below a few or tens of degrees Kelvin, depending on the impurity/matrix system), it is possible to detect the zero-phonon spectral lines (ZPL), that correspond to purely electronic transitions in impurity molecules. Parameters of ZPL such as frequency, intensity, width, time dynamics, etc. are very sensitive to the parameters of the local environment of the corresponding chromophore molecule, which makes it possible to use OM as spectral probes to obtain data on the structure and internal dynamics of solids, as well as other local values of various characteristics of the medium.

In most cases, at the temperatures up to dozens of Kelvin, ZPL broadening due to the quadratic electron-phonon interaction is discussed. In general, this contribution can be described in terms of the theory of I.S. Osad'ko [4]:

$$\Delta\Gamma_{e-phon}(T)=\frac{1}{4\pi}\int_0^\infty d\omega\cdot\ln\left[1+\frac{4n(\omega)[n(\omega)+1]W^2\cdot\Phi_{Ph}^2(\omega)}{[1-W\cdot\Omega(\omega)]^2+W^2\cdot\Phi_{Ph}^2(\omega)}\right], \quad (1)$$

where $\Omega(\omega)=\frac{2}{\pi}\int_0^\infty d\nu\cdot\Phi_{Ph}^2(\nu)P\left(\frac{\nu}{\omega^2-\nu^2}\right)$; $n(\omega)=\frac{1}{\exp(\hbar\omega/kT)-1}$ and $\Phi_{ph}(\nu)$ is the density of vibrational states in the matrix. In the case when the long- wave approximation and the weak

* Corresponding author: anikushina@isan.troitsk.ru

coupling approximation are applicable for the impurity-matrix system under study, and also if the density of vibrational states can be described by the Lorentz function corresponding to the effective (quasi-localized) low-frequency vibrational mode (LFM), expression (1) is simplified to

$$\Delta\Gamma_{e-phon}(T) = B\frac{\exp(-h\nu/kT)}{[1-\exp(-h\nu/kT)]^2} \;. \qquad (2)$$

For the most diverse impurity systems (glasses, polymers, molecular crystals), the temperature dependences of ZPL width demonstrated the validity of the indicated approximations for the study of bulk (macroscopic) samples. Such an analysis can also be carried out at the level of single molecules (SM) interacting with their local environment, which makes it possible to measure the value of the frequency (energy) of the LFM and the constant of quadratic electron-phonon coupling [5, 6].

Detection of fluorescent images allows to determine SM spatial coordinates with subdiffraction accuracy by computer analysis of image of single molecule taking into account the point-spread function [2, 3].

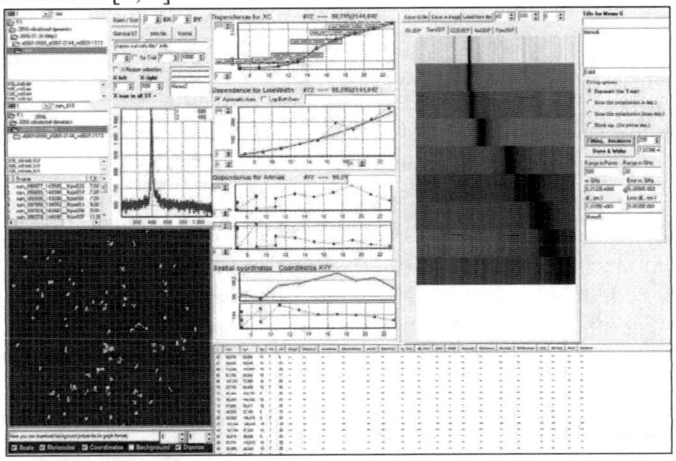

Fig. 1. Screen-shot of the special home-built computer program for combined analysis of temperature dependences of the fluorescence spectra and images of single molecules.

Thus, on the one hand, the direct experimental verification of the applicability of various approximations (leading to simplification of the Eq. (1)) on the level of a single molecule and its local environment and on the other hand was produced the mapping of LFM energy values in the sample.

T.A. acknowledges the grant of RFBR for young researchers (16-32-00841). The methods of statistical processing of SMSM data were developed under support of Russian Science Foundation (proj. No 14-12-01415).

References

[1] M. Orrit et al., Faraday Discussions **184**, 275 (2015)
[2] A.V. Naumov, Physics-Uspekhi **56**, 605 (2013).
[3] A.V. Naumov, I.Y. Eremchev, A.A. Gorshelev, Eur. Phys. J. D **68**, 348 (2014)
[4] I.S. Osad'ko, *Selective Spectroscopy of Single Molecules* (Springer, 2003)
[5] A.V. Naumov, Yu.G. Vainer, L. Kador, Phys. Rev. B **79**, 132201 (2009)
[6] Yu.G. Vainer, A.V. Naumov, L. Kador, Phys. Rev. B **77**, 224202 (2008)

Studying frequency correlations of photon pairs generated via spontaneous four-wave mixing in a microring resonator

Ilya Chuprina[1,2*], and *Alexey* Kalachev[1,2]

[1]Kazan E. K. Zavoisky Physical-Technical Institute, 10/7 Sibirsky tract Str., Kazan, 420029 Russia
[2]Kazan Federal University, 420008 Kazan, Russia

Abstract. Spontaneous four-wave mixing (SFWM) in a microring resonator is considered theoretically. The joint spectral amplitude and Schmidt number of generated biphoton field are calculated numerically for different pump pulses. Optimal pulse shapes that provide maximum efficiency of SFWM and minimum frequency correlations between signal and idler photons are found.

The nonlinear process of spontaneous four-wave mixing (SFWM) is a promising way to build heralded single-photon and entangled two-photon sources, which are essential elements in a number of photonic quantum technologies [1]. Integration of such sources on-chip is considered as the most promising approach to making them compatible with existing photonic architectures and scaling the optical schemes [2]. In doing so, to reduce the required pump power from the milliwatt to the microwatt regime, microring resonators prove to be very useful [3], wherein the rate of SFWM increases as the mode volume decreases, and the larger free-spectral range makes it easier to extract the emitted photons. It has been shown recently that short and broadband pump pulses allow reducing or even eliminating frequency correlation between the emitted photons [4, 5], which is crucial for preparing pure heralded single-photon states. In the present work, we study the dependence of frequency correlation and efficiency of photon pair generation on the pump pulse waveform.

The SFWM process involves sending a pump laser beam into a third-order nonlinear material so that the nonlinear interaction occasionally leads to the annihilation of two pump photons and creation of two photons in other modes (referred to as signal and idler) under conditions of energy conservation and phase matching. The quantum mechanical two-photon state is described by the joint spectral amplitude (JSA) $F(\omega_s, \omega_i)$, which is the probability amplitude of measuring a pair of photons at frequencies ω_s and ω_i. In possession of JSA we can calculate the Schmidt number K, which describes frequency correlation of the two-photon state. In the present work, we consider SFWM in a silicon-nitride racetrack resonator on silicon dioxide substrate and follow the theory developed in [6]. The joint spectral amplitude of biphoton field is calculated numerically for different pump pulses such as Gaussian, hyperbolic secant, bilateral exponential and rising exponential pulse. In doing so, we consider different values of ratio between the pump resonance linewidth $\delta\omega_p$ and

* Corresponding author: ilya.chuprina@gmail.com

signal/idler resonance linewidth $\delta\omega_{s,i}$. Some results of the calculation are shown in Fig. 1, where the joint spectral intensity (JSI) $|F(\omega_s, \omega_i)|^2$ and the corresponding Schmidt number are displayed for Gaussian and rising exponential pulses.

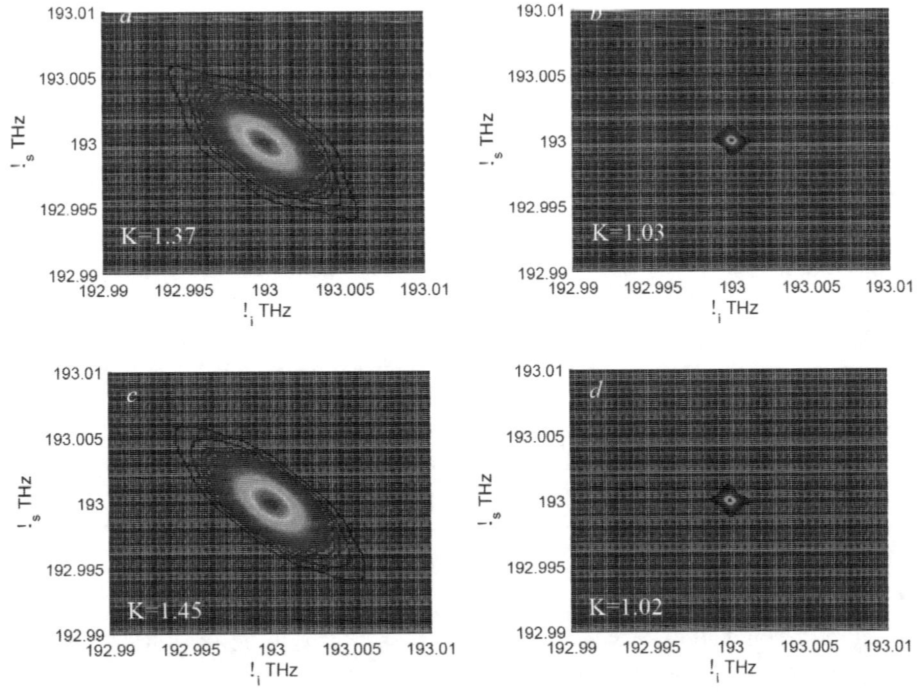

Fig. 1. The calculated JSI and Schmidt number of the biphoton field for rising exponent pump pulse (*a*), (*b*) and for Gaussian pump pulse (*c*), (*d*), and for different pump resonance linewidth: $\delta\omega_p = \delta\omega_{s,i}$ (a), (c) and $\delta\omega_p = 10\delta\omega_{s,i}$ (*b*), (*d*). In all cases, the temporal duration of the pump pulses is optimized to provide maximum efficiency.

According to the obtained results, among considered pulse shapes, Gaussian pulses provide maximum efficiency for any value of $\delta\omega_{s,i}/\delta\omega_p$. When $\delta\omega_p = \delta\omega_{s,i}$, the rising exponential pulse, which is the optimal for excitation of a single cavity mode, is less effective but provides the minimum Schmidt number. However, in the case of $\delta\omega_p \gg \delta\omega_{s,i}$, which is necessary for generating factorable two-photon states [4, 5], minimum Schmidt number is provided by bilateral exponential pulses.

This work was supported by the Russian Science Foundation (project No. 16-12-00045).

References

[1] J.L. O' Brien, A. Furusawa, J. Vuckovic, Nature Photonics **3**, 687 (2009)

[2] S. Tanzili, A. Martin, F. Kaiser et al., Las. Photon. Rev. **6**, 115143 (2012)

[3] W. Bogaerts, P. De Heyn, T. Van Vaerenbergh et al., Las. Photon. Rev. **6**, 47 (2011)

[4] L.G. Helt, Z. Yang, M. Liscidini, J.E. Sipe, Opt. Lett. **35**, 3006 (2010)

[5] Z. Vernon, M. Menotti, C.C. Tison et al., Opt. Lett. **42**, 3638 (2017)

[6] I.N. Chuprina, P.P. An, E.G. Zubkova et al. Quant. Electron. **47**(10), 887 (2017)

EPJ Web of Conferences **161**, 03003 (2017) DOI: 10.1051/epjconf/201716103003
PECS-2017

Data processing approaches in adopting single point-like emitter spectromicroscopy for mapping material characteristics of solid media

Alina Golovanova[1,2,*]

[1]Institute for Spectroscopy RAS, 108840 Troitsk, Moscow, Russia
[2]Moscow State Pedagogical University, 119991 Moscow, Russia

Abstract. We report an algorithm for building topogram of dielectric properties of a thin film by data of spectromicroscopy of single emitters.

Spectromicroscopy of single quantum emitters has been recognized a powerful tool to study parameters of solid transparent mediums [1,2]. For example, in recent works it was proposed to use single fluorescent molecules as spectral probe by which it is possible to determine the effective local value of the refractive index n [3].

Processing of experimental data obtained from the spectra and the fluorescence images of single molecules (SMs) in solid transparent matrices is known to have several problems which can be divided between physical-chemical and computational [4,5]. The first problem includes the presence of erroneous events, presents of other molecules, defects. The second problem includes huge or very small amount of data, hardware capabilities, etc. The problem considered in this work was a limited data and a significantly inhomogeneous spatial distribution of SMs over the plane of the sample. Note that these and related problems relates also to the spectromicroscopy of single semiconductor quantum dots [6]. The task of this study was to develop an algorithm for the correct transition from a discrete picture to a continuous distribution SMs data, to estimate the quality of such a transition and to construct a topogram based on SMs distributed somehow over the sample's plane.

The algorithm for building a topogram requires a certain data format. In our case, in the original data set, the local density of SMs in the sample plane fluctuates significantly, which makes direct averaging difficult. In order to eliminate this problem, the following algorithm was developed. The data plane is split by a lattice (grid) with a step r. Each node in the grid becomes the center of a circle with a fixed radius R. Each circle collects a number of SMs, while the minimum number of collected SMs depends on the selected R. The averaged data in each circle aria were used to give the average value of the effective refractive index of the film according to [7,8]. Each obtained value is attached to node with the coordinate XY. We have two key parameters: a grid step r and a radius R. The radius must be quite big to include the minimum quantity of SMs and, at the same time, be limited so that averaging does not take place over all available SMs. If we choice too large R, we loss information about local changes in the structure. Consequently, with an optimal

* Corresponding author: golovanova@isan.troitsk.ru

selection of these parameters, a more informative map of the values of n can be constructed.

Such an analysis was carried out on the example the generated model array of effective refractive index values distributed randomly in the surface of the sample (with size 50 to 50 arbitrary units) (Fig.1.(a)). The values of n are generated in accordance with the normal distribution law (the mathematical expectation is 1.5, the standard deviation is 0.1). As a result of processing obtained array we is constructed a corresponding topogram (Fig.1.(b)).

For the transition from the discrete distribution of n to the continuous one we used a standard interpolation packet based on radial basis functions (RBF) [9].

The algorithm was tested on the data obtained for SMs of an organic dye in a thin polycrystalline film.

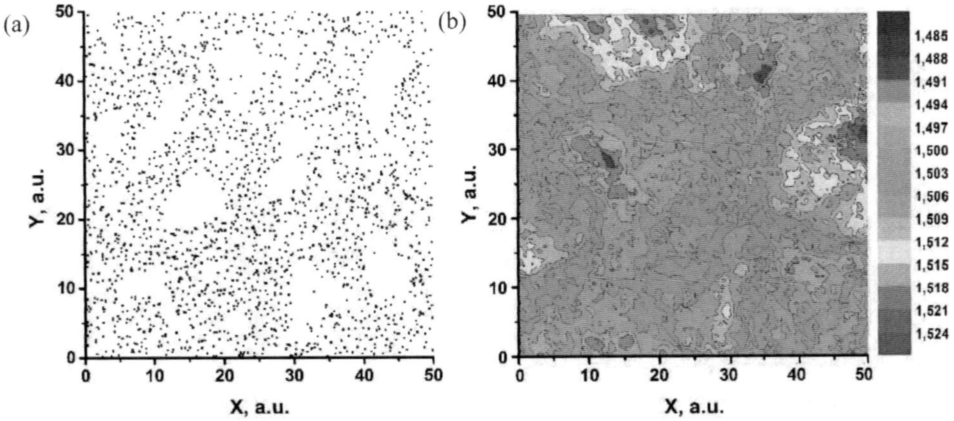

Fig. 1. (a) Model data distributed randomly in the surface of the sample with size 50 to 50 arbitrary units; (b) Topography of model data. Color scale shows the distribution of effective refractive index n from 1.485 to 1.524 with step 0.003.

Thus, the possibility of mapping the material characteristics (in this case n) of a continuous medium based on the discrete effective local n values obtained with the use of point spectral probes (SMs) randomly distributed in the plane of a thin-film sample was shown. It is important to note that the developed algorithm can be used even if the local data is missing in some regions of the sample.

This work was supported by Grant of the Russian Science Foundation (# 14-12-01415). Author is the member of the Leading Russian Scientific School (НШ–7035.2016.2).

References

[1] A.V. Naumov, Physics-Uspekhi **56**, 605 (2013)

[2] A.V. Naumov, I.Y. Eremchev, A.A. Gorshelev, Eur. Phys. J. D **68**, 348 (2014)

[3] T.A. Anikushina, M.G. Gladush, A.A. Gorsheleva, A.V. Naumov, Faraday Discuss. **184**, 263 (2015)

[4] A.V. Naumov, A.A. Gorshelev, Y.G. Vainer, L. Kador, J. Kohler, Ang. Chemie. Int. ed. **51**, 182 (2009)

[5] A.V. Naumov, A.A. Gorshelev, Y.G. Vainer, L. Kador, J. Kohler, Phys. Chem. Chem. Phys. **13**, 1734 (2011)

[6] A.L. Shchukina, I.Y. Eremchev, A.V. Naumov, Phys. Rev. E **92**, 032102 (2015)

[7] D.V. Kuznetsov, Vl.K. Roerich, M.G. Gladush, JETP **140**(4), 742 (2011)

[8] M.G. Gladush, EPJ Web Conf. **132,** 03007 (2017)

[9] M.D. Buhmann, Acta Numerica **9**, 1 (2000)

EPJ Web of Conferences **161**, 03004 (2017) DOI: 10.1051/epjconf/201716103004
PECS-2017

Determination of QW laser diode degradation based on the emission spectrum

Vladimir Bliznyuk[1,*], *Olga* Koval[1], *Elena* Vinogradova[1], and *Alexey* Rzhanov[2]

[1]Moscow Power Engineering Institute, 17 Krasnokazarmennaya str., Moscow, 111250 Russia
[2]Lomonosov Moscow State University, Physics Department, Leninskie Gory, Moscow 119991, Russia

Abstract. The possibility of laser diodes degradation control by monitoring of their spectrum is shown. For red and infra-red laser diodes, the time dependence of the radiation spectrum width was obtained.

Laser diodes of small and medium power (1-100 mW) are widely used as radiation sources in various complexes in metrology, medicine, optical transmission and information recording, and in many others. An important problem in the operation of these devices is the prediction of their service life and the determination of the operating time through which their substitution is required. To solve this problem, it is necessary to know which physical processes lead to the degradation of laser diodes. A study of degradation expressions and forms, carried out in [1-2], showed that its main cause is the gradual change in stresses in quantum-well semiconductor layers due to the difference in the parameters of the crystal lattices forming the active and emitter regions. Despite the efforts made by technologists to improve the quality of the layers by selecting suitable combinations of materials and improving their structure due to technological developments [3-4], defect formation processes in quantum-well active layers are difficult to predict. To predict laser operating time by the initial parameters, such as the output power, efficiency, wavelength and the degree of polarization of radiation specified in the instrument's passport is appeared to be impossible. Accelerated testing of the laser diode, conducted selectively, allow to detect changes in these parameters during the operation of the device, which, when generalizing the results of such measurements, makes it possible to predict the service life of a set of lasers. Therefore, it is very important for consumers to obtain techniques based on such a generalization. At present a number of methods for diagnosing the state of laser diodes have been developed. Despite that many of them are of interest from the point of view of physical or mathematical models, the question arises of their practical applicability. In our opinion, it is possible to obtain sufficiently complete information on the change in the state of the laser diode as the operating time increases, by its spectral characteristics analysis.

In this paper, we studied the spectral characteristics of two sets of eight and four laser diodes KLM650-5-5 and KLM980-5-15, respectively, generating red (650 nm middle wavelength) and infrared (980 nm) light. Laser modules were studied during the year and were exploited for 15 hours weekly. Every week during the single measurement, they were subjected to a three-fold heating to 70°C and cooling to -10°C. These temperatures were

* Corresponding author: 40595141@mail.ru

DOI: 10.1051/epjconf/201716103004

maintained in the climatic chamber for two hours each time, and the emission spectra of the lasers were measured the next day at room temperature and at an increasing pump current corresponding to a constant given value of the laser radiation power. It was recorded that the width of the spectral line of generation of the test laser diodes began to increase monotonically after the second month of investigations. It was found that the linc broadening speed in the interval between two to seven months' lasers remains constant, and from seven to 10 months it gradually decreases, up to zero at the finish of the annual test.

In Fig. 1 and Fig. 2 the spectra of the tested lasers before and after the year of operating in the regimes described above are shown.

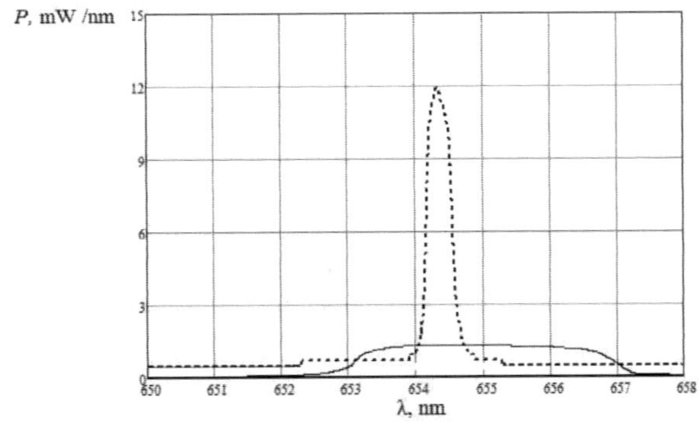

Fig. 1. Spectral characteristics of the KLM-D650-5-5 diode, measured at the beginning of the work (dotted curve), and after a year of operation (solid curve).

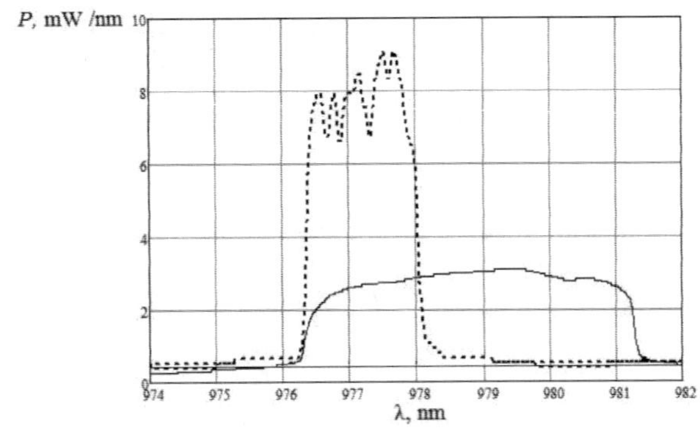

Fig. 2. Spectral characteristics of the KLM980-5-15 diode, measured at the beginning of the work (dotted curve), and after a year of operation (solid curve).

References

[1] V.V. Bliznyuk, M.A. Brit, I.S. Gadaev et al., Bull. RAS: Physics **79**, 12 (2015)

[2] O.I. Koval, A.G. Rzhanov, G.A. Solovyev, Phys. of Wave Phenom. **21**, 4 (2013)

[3] V.P. Duraev, A.A. Marmalyuk, A.A. Padalitsa et al., Quant. Electron. **38**, 2 (2008)

[4] S.O. Slipchenko, A.A.Podoskin, D.A. Vinokurov et al., Semiconductors **47**, 1079 (2013)

Multichannel all-optical switch based on a thin slab of resonant two-level emitters

Ramil Malikov[1,*], and *Victor* Malyshev[2]

[1]M. Akmullah State Pedagogical University of Bashkortostan, 450000 Ufa, Russia
[2]Zernike Institute for Advanced Materials, University of Groningen, Nijenborgh 4, 9747 AG Groningen, The Netherlands

Abstract. We discuss the possibility of using a thin layer of inhomogeneously broadened resonant emitters as a multichannel all-optical switch. Switching time from the lower stable branch of the system's bistable characteristics to the upper one and vice versa, which determines the speed of operation of a bistable device, is studied.

Bistability of an optical system is an extraordinary nonlinear effect, the essence of which is that the system has two stable states of transmission at the same value of the external field [1]. This phenomenon is of ongoing interest for optical technologies, in particular, for creation of fully optical switches, optical transistors, and optical memory.

In this work, we model the optical response of a thin resonant slab (TRS) of thickness L with inhomogeneously broadened two-level emitters of density N_0. As is well known, for such systems, the local field correction (LFC) to the field acting on an emitter is of importance [2,3]. The basis of our model consists of equations for the density matrix elements ρ_{ab} (a,b = 1,2) of a two-level emitter and the field E inside the slab, which in the rotating wave approximation reads [2,3]

$$\dot{R} = (i\Delta - \Gamma_2)R - \Omega Z\,, \tag{1}$$

$$\dot{Z} = \frac{1}{2}(\Omega^* R + \Omega R^*) - \Gamma_1(1+Z)\,, \tag{2}$$

$$\Omega = \Omega_0 + (i\Delta_L - \Gamma_R)\langle R\rangle, \tag{3}$$

$$\langle R\rangle = \int R(x,t,\Delta)G(\Delta)d\Delta\,. \tag{4}$$

Here, R is the amplitude of the off-diagonal density matrix element $\rho_{ab} = -(i/2)Re^{-i\omega t}$; $Z = \rho_{22} - \rho_{11}$; Γ_1 and Γ_2 are the relaxation constants of the population and coherence, respectively; $\Omega_0 = dE_0/\hbar$ and $\Omega = dE/\hbar$ are the external field E_0 and the field E inside the slab in frequency units; d is the transition dipole moment of the emitter; $\hbar$ is the reduced Plank constant; $\Delta = \omega - \omega_{21}$ is the detuning away from resonance of an isolated emitter, ω_{21}; $\Gamma = 2\pi d^2 N_0 kL/\hbar$ is the superradiant constant [2,3] and $\Delta_L = 4\pi d^2 N_0/(3\hbar)$ is the LFC [2,3]; $G(\Delta)$ is the inhomogeneous contour, in what follows, a Lorentzian of width Γ_2^*.

* Corresponding author: rfmalikov@mail.ru

 DOI: 10.1051/epjconf/201716103005

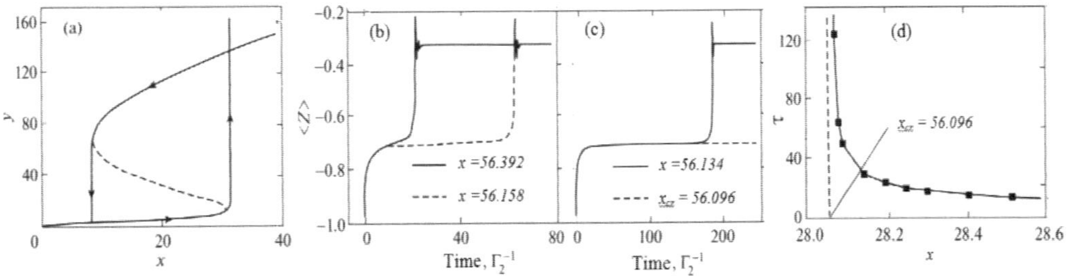

Fig. 1. (a) – Optical hysteresis of the field intensity $y = |\Omega|^2/(\Gamma_1\Gamma_2)$ inside the slab, obtained Under adiabatically sweeping the external field intensity $x = \Omega_0^2/(\Gamma_1\Gamma_2)$ up and down (shown by arrows). The parameters of calculations are : $\Delta_L = 20\Gamma_2$, $kL = 0.1$, $\Delta_0 = -5\Gamma_2$, $\Gamma_1 = 2\Gamma_2$, $\Gamma_2^* = 0.5\Gamma_2$. (b) and (c) – Time evolution of the integrated population difference $\langle Z \rangle$ after an abrupt switching of the external field intensity x from zero to a value that exceeds the critical intensity $x_{cr} = (\Omega_0^{cr})^2/(\Gamma_1\Gamma_2) = 56.096$. The parameters of calculations are the same as in (a), except $\Gamma_2^* = 2\Gamma_2$. (d) – Intensity dependence of the switching time τ.

Examples of numerical calculations of the optical response of the TRS are presented in Fig. 1. Plot (a) shows the optical hysteresis of the system: a sudden switching from one stable state to the other one and back under the adiabatically sweeping the driving field Ω_0 up and down (shown by arrows), occurring at different magnitudes of Ω_0. So the system is bistable. Plots (b) and (c) display the dynamics of switching at the high-intensity switching point $x_{cr} = (\Omega_0^{cr})^2/(\Gamma_1\Gamma_2) = 56.096$ (for the parameters specified in the figure caption). Plot (d) shows the intensity dependence of the switching time τ. Conspicuous is the fact that the dynamics of switching reveals a critical slowing down around the high-intensity switching point x_{cr}. Switching down always occurs during a time interval on the order of Γ_1^{-1}.

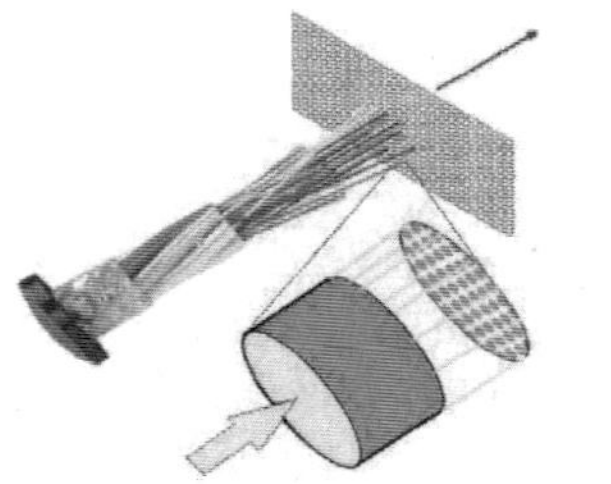

Fig. 2. Schematics of a multichannel all-optical switch (see details in the text).

Bistable properties of a thin resonant slab allow one to view this system as an all-optical switch. Apparently, different regions of the TRS are switched independently of each other. Then, this system may operate as a multichannel all-optical switch that is of importance from the viewpoint of optical information processing and optical computing. As an implementation of a device of such a type, an array of dielectric optical waveguides coupled with a TRS could be proposed, as illustrated in Fig. 2.

R. F. M. acknowledges the M. Akmullah State Pedagogical University of Bashkortostan for a support.

References

[1] S. L. McCall, Phys. Rev. A **9**, 1515 (1974)
[2] M.G. Benedict, A.I. Zaitsev, V.A. Malyshev et al., Opt. Spectrosc. **68**, 473 (1990)
[3] R.F. Malikov and V.A. Malyshev, Opt. Spectrosc. **122**, 955 (2017)

Technology of obtaining fine textured films for femtosecond instrumentation based on photon echo

R.W. Belyaev, A.U. Bakhadurov, N.S. Vashourin, K.Sh. Gazizov, A.N. Musanov, D.A. Nikitin, I.I. Popov*, A.V. Moroz, S.A. Stepanov, and N.I. Sushentsov

Volga State Technological University, 3 Lenin sq., Yoshkar-Ola, 424001 Russia

Abstract. This paper presents the technological approaches to the creation of thin textured films by the method of magnetron sputtering with parameters ensuring excitation of detected photon echo signals in them and the adjusted metrological characteristics of devices on a photon echo.

The purpose of this report is to develop and study the technology of obtaining thin textured films for femtosecond instrumentation based on photon echo. The physical foundations of the developed instruments rely on the excitation and use of the properties of the photon echo on exciton and trion states localized on surface defects of thin films. The task of technology is to reproduce systematically the thickness of thin films and the parameters of surface defects based on magnetron sputtering of these films. The parameters of technological regimes of magnetron sputtering of films which affect the reproduction of surface defects and the thickness of these films are reported in the work. Different variants of obtaining single-layer and multilayered textured thin films obtained under various magnetron sputtering regimes are given. Thin structured films obtained by magnetron sputtering methods have a fibrous structure. The Roentgen-amorphous phase fills the gap between the filaments. The surface defects were formed on the surface of individual fibers on which the excited exciton states were localized. The exciton states are described by the wave function typical for the superposition state.

Nearly located defects had closely located exciton lines with a lower binding energy than that of the surface particles of the fibers that do not have defects. They formed a cluster of neighboring particles with surface defects surrounded by an energy barrier of particles that do not have defects (Fig. 1). Because of the localization on the defects, the excitons are strongly bound to the phonons of the lattice and in the absorption and recombination spectra of excitons there is a phonon wing proportional to the density of the phonon states. This leads to a strong inhomogeneous broadening of the exciton lines.

At the interface between the crystalline and amorphous phase, changing various parameters of the technological process, such as the substrate temperature, the sputtering time, the composition of the gas mixture (Fig. 2), the number and composition of layers in

* Corresponding author: Popov@volgatech.net

the multilayer film (Fig. 3), and others, there comes the possibility of the controlled introduction of the surface defects. In this paper, methods and means for controlling the production technology of films are proposed, which make it possible to obtain films with a given configuration of surface defects. The films thus obtained are used to demonstrate the physical basis of the principle of constructing a device based on the photon echo.

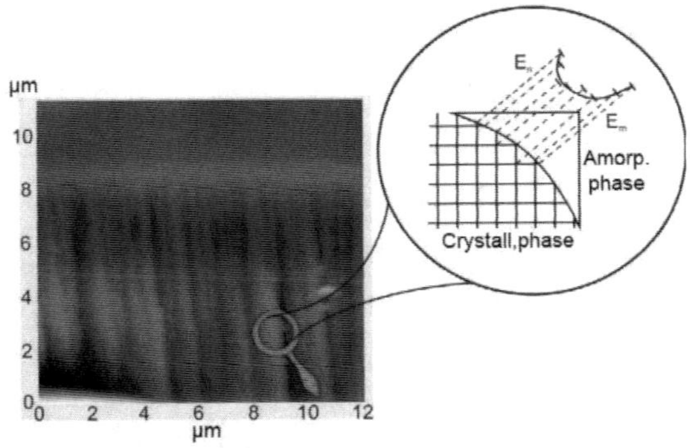

Fig. 1. The image of the cleaved film

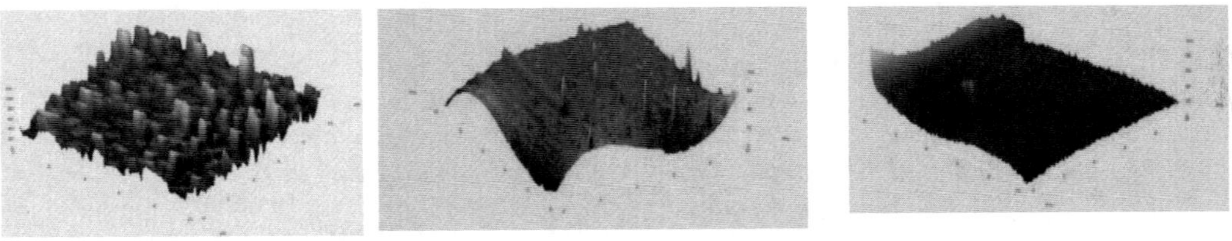

Fig. 2. The shapes of the surface and the protruding above it crystalline fibers of the films, when various gases except the sputtered target material were introduced into the chamber, at the moment when a film of ZnO (a), Si (B), Si (P) was formed.

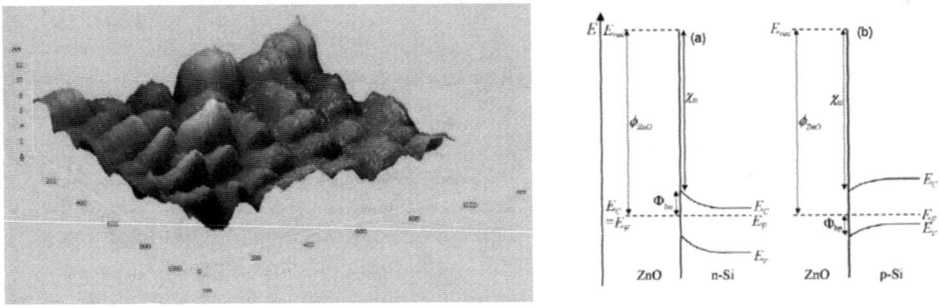

Fig. 3. The shape of the surface and the protruding above it crystal fibers (a) and the band structure of the ZnO heterojunction with Si n-type and Si p-type for a three-layer film [1].

References

[1] V. Quemener, M. Alnes, L. Vines, et al., Journal of Physics D **45**(31), 315101 (2012)

Long-lived femtosecond photon echo in thin films on localized exciton states at room temperature

N.S. Vashurin[1], V.O. Kompanets[2], I.I. Popov[1,*], S.E. Putilin[3], and S.V. Chekalin[2]

[1]Volga State Technological University, 3 Lenin sq., Yoshkar-Ola, 424001 Russia
[2]Institute of Spectroscopy of the Russian Academy of Sciences, 5 Fizicheskaya str, Moscow, Troitsk, 108840 Russia
[3]St. Petersburg National Research University of Information Technologies, Mechanics and Optics, 49 Kronverksky Prospect, St. Petersburg, 197101 Russia

Abstract. Long-lived femtosecond photon echo was excited at room temperature in a two-photon regime on the charged exciton states (trion type) localized on the surface defects of a three-layer textured ZnO/Si(B)/Si(P) film with the thickness of each layer at 100 nm in a longitudinal uniform magnetic field. The irreversible longitudinal relaxation time T_1 for the long-lived photon echo regime was of the order of 40 ps, whereas when measured with the help of the stimulated photon echo generated in the single-photon mode, the value was found not to exceed 600 fs.

The paper reports an experimental observation of the regime of a long-lived femtosecond photon echo at room temperature on exciton states localized on surface defects of a three-layer textured ZnO/Si(B)/Si(P) with a thickness of each layer of 100 nm. The excited film was exposed to a magnetic field applied in the direction that was longitudinal with respect to the propagation direction of the laser excitation. This textured film was obtained by magnetron sputtering. The decrease in the intensity of a femtosecond stimulated photon echo with increasing time interval between the second and third exciting impulses was recorded in the presence and absence of a uniform longitudinal magnetic field. The time interval between the first and second exciting pulses was chosen to be zero. The duration of the exciting impulses was of the order of 40 fs. The intensity of the applied magnetic field was 150 mT. The angle between the directions of propagation of the exciting pulses was about 9 degrees. The direction of recording the echo signal corresponded to the conditions of spatial synchronism for the stimulated photon echo excited in the two-photon regime.

In the presence of a longitudinal homogeneous magnetic field of the echo, a long-lived photon echo effect was observed in a three-layer film (Fig. 1), in single-layer films it did not. When an echo signal corresponding to the synchronism conditions for four-wave mixing and the stimulated photon echo excited in the single-photon mode was observed, the usual femtosecond stimulated photon echo (SPE) was recorded. The time of irreversible

* Corresponding author: Popov@volgatech.net

longitudinal relaxation T_1 for the regime of the long-lived photon echo was about 40 ps, whereas when it was measured with the help of the SPE generated in single-photon mode, a value of no more than 600 fs was obtained.

The long-lived regime is realized, thanks to the presence of the spectral line, of trionic states formed not only in single-photon, but also in two-photon excitation regimes. Metastable (long-lived) quantum levels were excited in the two-photon regime.

The role of the magnetic field here has two functions. The first function, analogous to that shown by the authors of the work [1], is associated with periodic precession with the Larmor frequency of the magnetic moment of these two states, which ensures the transfer of the excitation of the usual quantum level to a long-lived one.

The second function of the magnetic field is associated with the rotation of the magnetic moment of the pseudo-electric dipoles arising during the formation of the photon echo around the direction of the longitudinally applied homogeneous magnetic field. In this case, the Larmor precession, acting on pseudoelectric dipoles, affects the phase precession rate of these dipoles. This contribution differs in sign for the upper (right-handed) and lower (left-handed) components of the hyperfine structure of the excited level. Which ultimately leads to a non-Faraday rotation of the polarization vector of the echo signal.

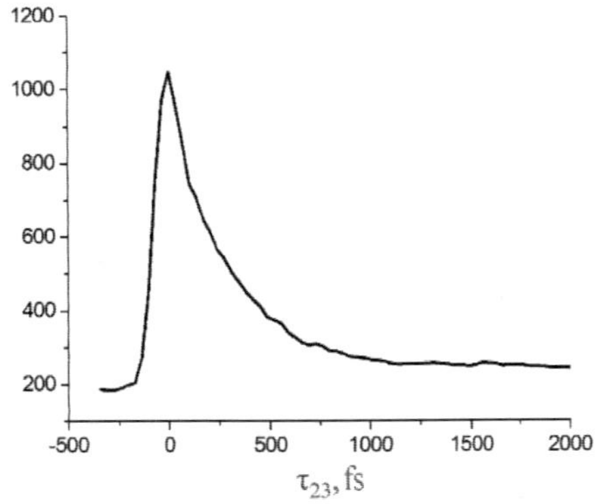

Fig. 1. Dependence of the intensity of the photon echo excited in the two-photon mode in the presence of a longitudinal homogeneous field on localized exciton states (such as the trion), on the value of the time interval between the second and third exciting impulses.

References

[1] L. Langer, S.V. Poltavtsev, I.A. Yugova et al., Phys. Rev. Lett. **109**, 157403 (2012)
[2] L. Langer, S.V. Poltavtsev, I.A. Yugova et al., Nature Photon. **8**, 851 (2014)

EPJ Web of Conferences **161**, 03008 (2017) DOI: 10.1051/epjconf/201716103008
PECS-2017

On-line registration of femtosecond time intervals based on polarization properties of femtosecond stimulated photon echo generated on exciton states

A.U. Bakhodurov[1], N.S. Vashourin[1], E.A. Vinogradov[2], K.Sh. Gazizov[1], V.O. Kompanets[2], I.I. Popov[1,*], S.E. Putilin[3], and S.V. Chekalin[2]

[1]Volga State Technological University, 3 Lenin sq., Yoshkar-Ola, 424001 Russia
[2]Institute of Spectroscopy of the Russian Academy of Sciences, 5 Fizicheskaya str, Moscow, Troitsk, 108840 Russia
[3]St. Petersburg National Research University of Information Technologies, Mechanics and Optics, 49 Kronverksky Prospect, St. Petersburg, 197101 Russia

Abstract. This paper reflects the results of the research on the character of the dependence of the non-Faraday rotation of the femtosecond stimulated photon echo polarization plane on the time interval between the second and third exciting pulses, discretely varying from 180 to 900 fs in increments 180 fs. The time interval between the first and second pulses was equal to zero. The echo signal was formed at room temperature on exciton states localized on the surface defects of a thin three-layer textured ZnO/Si(P)/Si(B) film in the presence of a homogeneous magnetic field of 0.25 mT applied longitudinally to the optical excitation axis. The qualitative coincidence of the investigated dependence with the theoretical prediction of the investigated effect for gaseous medium is shown.

Stimulated photon echo (SPE) was excited at room temperature by three laser pulses on exciton states localized on nanoscale surface defects of a three-layer ZnO/Si(B)/Si(P) thin film (each layer had a thickness of 100 nm). The experiment was carried out both without a magnetic field, and in the presence of a longitudinal homogeneous magnetic field with an induction of 0.25 mT. The dependence of the intensity of the SFE signal on the angle of rotation of the prism relative to the plane of polarization of the exciting laser pulses was recorded. The parameters of the exciting pulses were as follows: τi - 60 fs; λvi = 790 - 830 nm; Evi> 30 μJ; Pv - 1 TW / cm2. The time interval between the first and second pulses was equal to zero, and between the second and third it changed discretely.

The difference in the polarization plane of a femtosecond SPE excited on exciton states localized on the surface defects of a thin three-layer textured ZnO/Si(P)/Si(B) film depending on the presence of a magnetic field is given in the following work [1]. The difference in the value of the angle of rotation of the plane of polarization for time intervals between 2 and 3 pulses, equal to 720 and 900 fs, is given in the work [2]. In this paper we present the results

* Corresponding author: popov@volgatech.net

of a study of the non-Faraday rotation of the plane of polarization of a femtosecond photon echo from the time interval between the second and third exciting pulses from 180 to 900 fs in steps of 180 fs.

The decrease in the angle of rotation of the PE polarization plane with the growth of the time interval τ23 qualitatively confirms the coincidence of the results obtained with the theoretical prediction [3] of the tangential dependence of this effect for gaseous medium on the value of the time interval τ23. The decrease in the angle of rotation of the PE polarization plane with the growth of the time interval τ23 qualitatively confirms the coincidence of the results obtained with the theoretical prediction [3] of the tangential dependence of this effect for gaseous media on the value of the time interval τ23. The difference between the results of this experiment and the theoretical predictions is due, first of all, to the influence on the process of formation of the SPE on the presence of the phonon wing exciton state in the resonant spectral line, which is absent on the gas spectral line. Therefore, the form of the tangentoid is distorted in comparison with the experiment in the gas, but the self-tangential character of the investigated dependence is preserved. The points obtained refer to the negative branch of the tangentoid. By detecting the angle of the non-Faraday rotation of the polarization plane of a femtosecond photon echo, it is possible to register on-line femtosecond time intervals in the range of values starting with a doubled duration of the laser pulses that excite the echo signal, to within several times smaller magnitude than today's existing on-line registration capabilities of the smallest time interval.

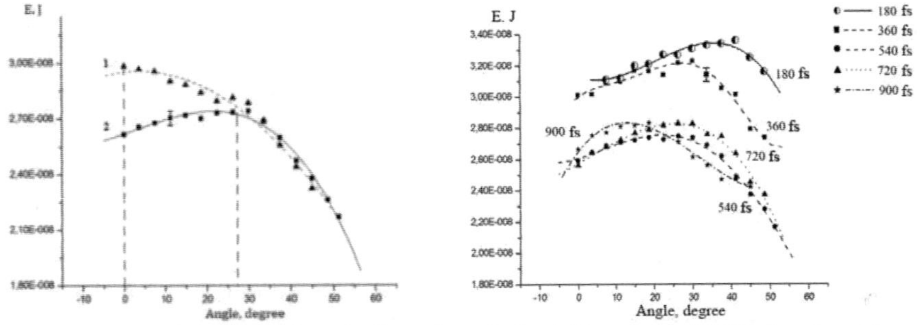

Fig. 1. The dependence of the intensity of a linearly polarized femtosecond SPE signal, on the angle of rotation of the prism relative to the plane of polarization of the exciting laser pulses for a time interval of τ_{23} equal to 720 fs: 1 - without a magnetic field; 2 - in the presence of a longitudinal homogeneous magnetic field of 0.25 mT (left). Dependence of the intensity of a linearly polarized femto-second SPE signal on the angle of rotation of the prism relative to the plane of polarization of exciting laser pulses for a time interval of τ23, discretely varying from 180 to 900 fs after 180 fs. The maximums of the curves for the five discrete values of τ23 were observed at the following prism angles φ: 1) τ_{23} = 900 fs; φ = 10 deg; 2) τ_{23} = 720 fs; φ = 27 deg; 3) τ_{23} = 540 fs; φ = 31 degrees; 4) τ_{23} = 360 fs; φ = 33 degrees; 5) τ_{23} = 180 fs; φ = 34 deg (right).

This work was partially financially supported by the Government of the Russian Federation, Project # 3.9041.2017/7.8

References

[1] N.S. Vashourin et al., Abstr. Of 9-th Int. Sci. Sch. "Science and Innovations-2014", Yoshkar-Ola: VolgaTech, 2014, P. 90-93 (in Russian)

[2] N.S. Vashurin, I.I. Popov and S.E. Putilin, Bull. RAS: Phys. **81**(5), 554 (2017)

[3] I.I. Popov et al., Bull. RAS: Phys. **81**(5), 586 (2017)

[4] I.V. Evseev, V.M. Ermachenko, V.A. Reshetov, Opt. Spectrosc. **52**(3), 444 (1982)

EPJ Web of Conferences **161**, 03009 (2017) DOI: 10.1051/epjconf/201716103009
PECS-2017

Polarization properties of superposition states of the quantum transition of molecular iodine vapor in a magnetic field and their possible application

K.Sh. Gazizov, and I.I. Popov*

Volga State Technological University, 3 Lenin sq., Yoshkar-Ola, 424001 Russia

Abstract. This paper considers the problem of the possibility of creating a technology for constructing a high-speed high-precision laser lidar based on the physics of the non-Faraday rotation effect of the polarization plane of the primary photon echo in molecular iodine vapor at room temperature. A method is proposed for increasing the range of distance measurements by combining the capabilities of an optical echo-range lidar and an optical time-pulse range lidar.

The possibility of creating a technology for constructing a high-precision fast-acting phase-pulse laser range finder (lidar) based on the effect of the non-Faraday rotation of the plane of polarization of the PE in molecular iodine pairs is presented in this paper [1, 2].

In the formation of the photon echo the process of creating equally polarized pseudoelectric dipoles happens in all particles simultaneously. After the end of the action of the short pulse, the process of dephasing the dipoles takes place, and then after the arrival of the next pulses - their phase shift. Which leads to the formation of a photon echo signal. At that, on the left-handed and dextrorotatory components of the hyperfine structure of the excited quantum level under the influence of the magnetic field, dephasing (phase shifting) of dipoles takes place at different phase velocities. This is the reason for the non-Faraday rotation of the polarization plane of the photon echo. The magnitude of this rotation depends on the strength of the magnetic field (the magnitude of splitting of the hyperfine components of the excited quantum level) and the time interval between the exciting pulses. It does not depend on the number of resonance particles, that is, on the length of the radiation path in this medium. The magnitude of the angle φ of rotation of the PE polarization plane is than 3 times bigger in exponent part than the Faraday rotation at the same value of the applied magnetic field intensity.

Using the dependence of this rotation on the time interval between excitation pulses, we propose a method of high-speed, high-precision distance measurement [3]. In such a laser rangefinder, the first exciting pulse is a reference pulse, and the second is a probing pulse. In this connection the recorded angle of rotation of the echo signal polarization plane contains information about the time of passage of the measured distance in both directions by the probing pulse. This time is recalculated into the measured distance.

The essence of the technology lies in the longitudinal application to the iodine vapor of the required intensity of a homogeneous magnetic field and the provision of longer relaxation time for iodine vapor than the sum of the durations of the exciting pulses and the time interval

* Corresponding author: popov@volgatech.net

separating them. The disadvantage of such a range finder is the limitations on the magnitude of the measured distance, determined by the magnitude of the irreversible transverse relaxation time T_2.

In this paper it is reported about the conjugation [4] of the above method of measuring distances with a time-pulse method. The first exciting laser pulse is applied to a resonant medium with a strictly fixed time delay in reference to the moment of departure to the distance of the probe pulse. The magnitude of this delay can be any, it corresponds to a strictly defined analytically added (or measured by means of a time-impulse range-finder) distance, to which the additional measurement made with a high precision by the photon echo is added. The sum of the delay time of the reference signal and the time interval measured by the echo-distance meter will be the recorded time for the probe pulse to travel two measured distances. The process of measuring the distance is made in 2 stages. At the first stage, the range finder operates in the time-impulse rangefinder mode. Measurement of the distance in this case occurs with an accuracy of one to tens of centimeters (depending on the duration of the exciting pulses). At the second stage, the range finder operates in the phase high-precision range finder mode. In this case, a time interval less than half the relaxation time T_2 is measured, which is about of 15 ns. This is possible due to the fact that the time of delivery of the first excitation pulse (reference pulse (OI) of the range finder) to the resonant medium is delayed by a time interval equal to the multiplication of the whole number of time intervals equal to half the time T_2 determined by the time-pulse range finder.

It is to this value that the reference pulse is delayed in order to reduce the time interval between the reference and probing (ZI) pulses of the range finder of relaxation to a value not exceeding the time of relaxation the medium T_2.

The angle φ is recalculated by the formula (1) into the value of the time interval τ_{12} separating the excitation pulses

$$\tau_{12} = \frac{2}{144.\times H} \times arctg \frac{-2 \pm \sqrt{4+3tg^2\varphi}}{tg\varphi} + 2\pi n, \qquad n \in Z,$$

where Z is the set of whole numbers (1).

The step of additional measuring the distance is chosen equal to 5 microns, which corresponds to the time of its passage by an optical pulse equal to 15 fs. The result of the measurement of the distance is summed up with the delay time of the first pulse, given with an accuracy of 15 fs, and of the additional distance measurement.

References

[1] I.I. Popov, I.S. Bikbov, I.V. Evseev, V.V. Samartsev, J. Appl. Spectr. **52**(5), 794 (1990)

[2] I.V. Evseev, V.M. Ermachenko, Opt. Spectrosc. **47**(6), 1139 (1979)

[3] K.Sh. Gazizov, A.M. Gladyshev, I.I. Popov, V.V. Samartsev, I.S. Bikbov, Patent IPC G01C3/00(2006/01) No. 23065 27 (November 7, 2005)

[4] K.Sh. Gazizov, I.I. Popov, and V.T. Sidorova, Physics of Wave Phenomena **20**(1), 14 (2012)

EPJ Web of Conferences **161**, 03010 (2017) DOI: 10.1051/epjconf/201716103010
PECS-2017

Generation of the rotating light fields under amplitude and phase distortions

Vladimir Volostnikov[1], *Evgeniy* Vorontsov[1], *Svetlana* Kotova[1,2], *Nikolay* Losevsky[1], *Darya* Prokopova[1,2,*], and *Sergey* Samagin[1]

[1]Lebedev Physical Institute, Samara, 443011, 221 Novo-Sadovaya Str., Samara, Russia
[2]Samara National Research University, 443086, 34, Moskovskoye shosse, Samara, Russia

Abstract. The work contains study of the influence of amplitude and phase distortions on the generation of two-lobe light fields with the intensity distribution rotating under propagation. Spiral beams, as well as light fields formed on their basis, are considered as particular examples of two-lobe light fields. The amplitude distortions result from the preset different intensity distributions of the illuminating beam, and phase distortions are determined by the primary aberrations.

The light fields with a rotating intensity distribution under propagation are the matter of interest in solving various applied problem. One of the bright examples of such problems is the use of two-lobe light fields (i.e. the field with two sharp intensity peaks) [1] for improvement of the longitudinal resolution of fluorescent optical microscopes. The ability of the intensity rotation during propagation is characteristic for spiral light beams [2]. To create these beams you need to set amplitude and phase of the light field that can be implemented by using of the holographic method [3]. However, the latter can not be always used due to the optical system characteristics and low energy efficiency. Therefore a compromise is often made in practice and light fields preserving the spatial structure of their intensity distribution within a limited area, are used instead of the spiral beams. These fields are formed by means of the phase diffractive masks. To calculate the mask phase profile, in our study we used the phase distributions of spiral beams as initial approximation.

Mostly in real optical systems both amplitude and phase distortions occur. For example, in problems of three-dimensional high resolution of the fluorescent microscopy, the intensity distribution of the beam illuminating the phase mask depends on the used micro objective and the ratio of the apertures of the mask and illuminating beam. The phase distortions can be caused by aberrations of the optical system, geometric parameters of the observing system and heterogeneity of the optical parameters of the medium where a light wave propagates. Usually the influence of phase distortions prevails over that of amplitude distortions.

In this paper we continue the research initiated by us earlier in our work [4]. For generation of the light fields, the phase masks based on the spiral beams were used, representing a superposition of Laguerre-Gauss modes. Primary aberrations in the representation of the Zernike polynomials are considered as the phase distortions. Numerical simulation and experiments were performed. A liquid crystal spatial light modulator was used

* Corresponding author: prokopovadv@gmail.com

 DOI: 10.1051/epjconf/201716103010

as tunable diffractive optical element for light fields formation. Aberrations of different amplitudes have been studied. In figure 1 the phase masks are shown, used for the formation of two-lobes fields with the rotating parameter -2, for the case of superposition of aberrations with the amplitude of λ/8. In figure 2 the results of numerical simulations intensity distribution in focal plane are shown for the case of aberrations of different amplitude. The value of the aberration in the system permissible for the work with this type of fields is estimated. The amplitude distortions were specified by different intensity distributions of the beam illuminating the phase mask.

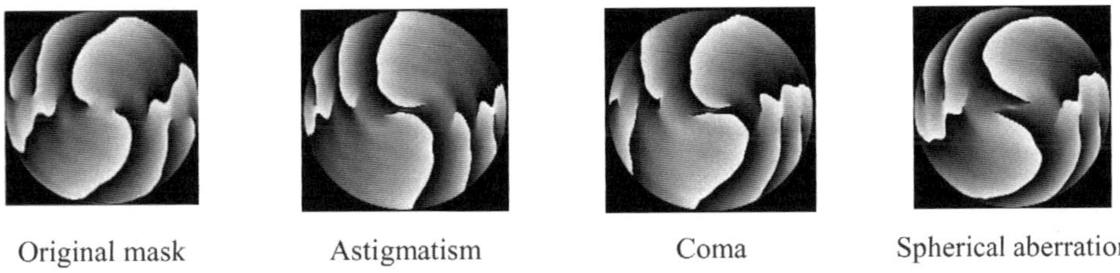

Fig. 1. Phase masks for generation of a two-lobe light field with rotating parameter -2 with superposed aberrations of amplitude λ/8

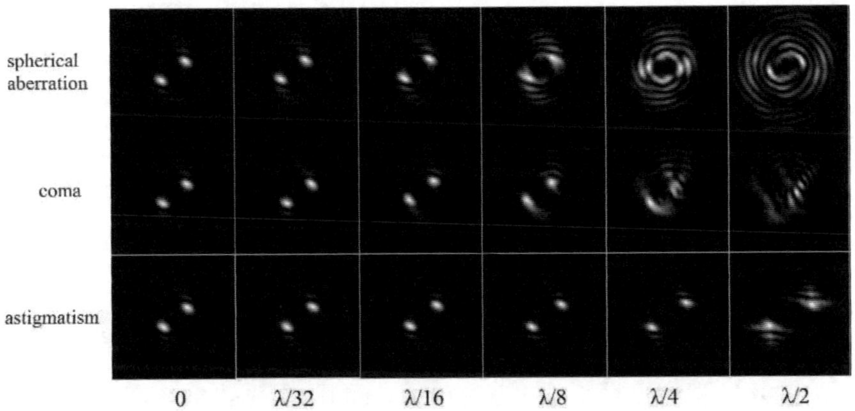

Fig. 2. Calculated intensity distribution in focal plane under the aberration of different amplitude for two-lobe light field with rotating parameter -2.

The work was supported by the Russian Foundation for Basic Research (Grant No. 16-29-11809) and the Program of Physical Science Department of the Russian Academy of Sciences.

References

[1] A.V. Naumov, I.Y. Eremchev, A.A. Gorshelev, Eur. Phys. J. D **68**, 348 (2014)

[2] E.G. Abramochkin, V.G. Volostnikov, *Modern Optics of Gaussian beams* (Moscow, Fizmatlit, 2010)

[3] K.N. Afanasiev, S.A. Kishkin, Izvestiya Samarskogo Nauchnogo Tsentra RAN **14**, 184 (2012) (in Russian)

[4] E.N. Vorontsov, N.N. Losevsky, D.V. Prokopova, E.V. Razueva, S.A. Samagin, Comp. Opt. **40**, 158 (2016)

DOI: 10.1051/epjconf/201716103011

Correlated signals of cluster superfluorescence under two- and three-photon excitation of CdSe/CdS nanostructures

Vitaly Samartsev, and *Tatiana* Mitrofanova*

Kazan E. K. Zavoisky Physical-Technical Institute, 10/7 Sibirsky tract str., Kazan, 420029 Russia

Abstract. The possibility and conditions for generation of the correlated signals of cluster superfluorescence (CSF) under two and three-quantum excitation of nanostructured samples CdSe/CdS by two crossed at the angle of 60° femtosecond beams of the Ti:Sapphire laser radiation are investigated. It is shown that intensity of the CSF signals is proportional to the cube of the number of clusters and that these collective signals are generated in mutually opposite directions $\mathbf{k}_1 - \mathbf{k}_2$ and $\mathbf{k}_2 - \mathbf{k}_1$, where $\mathbf{k}_1$, $\mathbf{k}_2$ are the wavevectors of the exciting pulses.

R. Dicke in his basic work [1] devoted to the optical superradiance (SR) also discussed the case of an incoherent pump when at the initial moment of time the radiators (atoms, molecules or clusters) are in the excited state and have no macroscopic polarization. It was shown that under conditions when the linear dimensions of the system are much less than the wavelength λ of the radiated light such a system of the inverted atoms can spontaneously undergo transition to the ground-state during the time which is inversely proportional to the number of atoms N. This case of the spontaneous collective radiation was named in [2] as a superfluorescence (SF). The SF, as well as the SR, is a perfect example of a self-organization [3]. The collective character of photon emission in the SF is due to the indistinguishability of atoms in a system (with linear dimensions $L < \lambda$) interacting with a field. The SF signals appear to be delayed by the time $t_0 = \tau_c \ln N$, where $\tau_c = T_1 / N$ is the time of correlation self-formation, T_1 is the lifetime of the excited state, N is the number of radiators.

The samples studied in the present work are composites consisting of the CdSe/CdS nanoparticles (of core/shell type) dispersed into a PMMA polymer matrix. The average size of grown particles with the spherical or elliptical shape is equal to $5 \cdot 10^{-7}$ cm. The wavelength of the Ti:Sapphire laser is equal to $8 \cdot 10^{-5}$ cm with pulse duration of 40 fs. So, the area of excitation with the linear dimensions of $8 \cdot 10^{-5}$ cm can involve about 160 areas (clusters) emitting the SF signal. For the first time the term "cluster superfluorescence" appeared in paper [4] while studying the quadrupole superradiance [5, 6]. The authors of this paper constructed the generalized Dicke model and showed that the intensity I of the quadrupole cluster superradiation is proportional to N^3. In our case, the situation is more complicated because of the two- and three-photon modes of the excitation by crossed femtosecond laser

* Corresponding author: samartsev@kfti.knc.ru

beams. The presence of the angle of 60° between the wavevectors of the first $\mathbf{k}_1$ and the second $\mathbf{k}_2$ exciting pulses leads, according to [7], to the emitting of the coherent CSF signals in the mutually opposite directions $\mathbf{k} = \mathbf{k}_1 - \mathbf{k}_2$ and $\mathbf{k}' = \mathbf{k}_2 - \mathbf{k}_1$.

The peculiarities of the excitation of the coherent responses in the semiconductor CdSe/CdS nanostructures by the Ti:Sapphire were discussed in details in [8]. Earlier, in [9], it was experimentally shown that one needs the three-photon mode in order to excite the electrons of such structures by the radiation of this laser. It was established that both conduction bands of the sample are excited coherently. But then electrons from the upper conduction band decay nonradiatively to the lower one. This is the reason why the nonequilibrium population grating at the energy transition between the valence band and the upper conduction band (257 nm) disappears. But this grating which is important for a calculation of the correlated signals of CSF still exists at the energy transition between the valence band and the lower conduction band (534 nm). Although the excitation has a three-photon character, the presence of the nonradiative transition allows us to simplify the situation and consider the two-photon mode of an excitation. Calculations performed by the method presented in [4] (see also [3]) give us the following formula for the intensity of CSF signals:

$$I(\mathbf{k}_{\mathrm{CSF}}) \sim I_0(\mathbf{k}_{\mathrm{CSF}}) N^3 M A \lambda^2 \,\mathrm{th}^2\left(\frac{\hbar\omega}{2k_{\mathrm{B}}T}\right) e^{-t'/T_2},$$

where $I_0(\mathbf{k}_{\mathrm{CSF}})$ is the intensity of the spontaneous radiation of the isolated cluster in the direction $\mathbf{k}_{\mathrm{CSF}}$, N is the number of active clusters, M is the parameter of quadrupole radiation of a cluster, A is a parameter of two-photon electron absorption, ω is the carrier frequency of the Ti:Sapphire laser, k_{B} is the Boltzmann constant, T is the temperature of the sample, t' is the decay time of CSF signal, T_2 is the time of transverse nonreversible relaxation.

This work is supported by the Program of the Presidium of RAS "Actual problems of the low temperature physics" and by the RFBR grant no. 17-02-00701-a.

References

[1] R.H. Dicke, Phys. Rev. **93**, 99 (1954)

[2] R. Bonifacio, L.A. Lugiato, Phys. Rev. A **11**, 1507 (1975)

[3] A.A. Kalachev, V.V. Samartsev, *Coherent phenomena in optics* (Kazan, Kazan State University, 2003) (in Russian)

[4] A.A. Kalachev, V.V. Samartsev, Proc. SPIE **6181**, 6181112 (2005)

[5] U.Kh. Kopvillem, V.R. Nagibarov, V.V. Samartsev, Phys. Lett. **42**, 399 (1973)

[6] Yu.V. Naboikin, V.V. Samartsev, P.V. Zinoviev, N.B. Silaeva, *Coherent spectroscopy of molecular crystals* (Kiev, Naukova Dumka, 1986) (in Russian)

[7] A.V. Leontiev et al., Laser Phys. Lett. **9**, 654 (2012)

[8] V.V. Samartsev, T.G. Mitrofanova, Laser Phys. **26**, 125203 (2016)

[9] J.W.M. Chon, M. Gu, Appl. Phys. Lett. **84**, 4472 (2004)

EPJ Web of Conferences **161**, 03012 (2017) DOI: 10.1051/epjconf/201716103012
PECS-2017

Luminescence decay of Sm:LaF_3@LaF_3 core-shell crystalline nanoparticles

Elena Lukinova*, *Eduard* Madirov, *Maxim* Pudovkin, *Darya* Koryakovtseva, *Stella* Korableva, *Alexey* Nizamutdinov, and *Vadim* Semashko

Kazan Federal University, Institute of Physics, 18 Kremlevskaja str., Kazan, 420008 Russia

Abstract. Luminescent properties in visible spectral range of crystalline LaF_3 nanoparticles containing 1 % and 5 % of Sm^{3+} ions before and after microwave hydrothermal treatment and also Sm:LaF_3@LaF_3 core-shell nanoparticles were studied. The LaF3 surface layer resulted in lowering the luminescence decay time apparently due to eliminating surface quenching factors. This opens the way to manage luminescence and energy transfer properties of the material.

1 Introduction

Crystalline nanoparticles of LaF_3 compound, doped with various rare-earth ions are of great interest today for unique combination of optical and physical properties. Depending on luminescence and excitation spectral ranges of action these materials are prospective in bioimaging, lamp phosphors, quantum cutting, up-conversion, luminescent material for combined photodynamic therapy [1,2]. Composite materials based on these nanoparticles can be utilized as elements for quantum devices. Aim of this work is to study opportunities to manage luminescence decay properties of Sm^{3+} ions in LaF_3 nanoparticles by chemical approach.

2 Experimental

Crystalline Sm:LaF_3 nanoparticles were fabricated via co-precipitation method described earlier [3,4]. Nanoparticles of Sm:LaF_3 (C_{Sm} = at. 1 %, C_{Sm} = at. 5 %, samples N1 and N2 respectively) were used in experiments after washing and drying directly after synthesis. The hydrothermal treated (60 minutes exposure in the microwave oven) samples were prepared as it can improve crystallinity of the particles (sample N3). Also the approach of core-shell was tested. The sample of N2 was dispersed in distilled water by sonication treatment. Then 1.2 mmol of NH_4F was added into suspension. Then the solution was heated to 50 °C, and stirred for 15 min and the 0.4 mmol of $La(NO_3)_3$·was introduced to it. The prepared suspension was placed into the microwave oven for 60 minutes. The following procedures of washing and drying were the same as that was stated for previous samples.

* Corresponding author: elena.v.lukinova@gmail.com

DOI: 10.1051/epjconf/201716103012

3 Results and discussion

Luminescence spectra of all Sm:LaF_3 samples registered under 405 nm excitation contain well known lines of transitions from $^4G_{5/2}$ state to ^6H_J terms (see Fig. 1a).

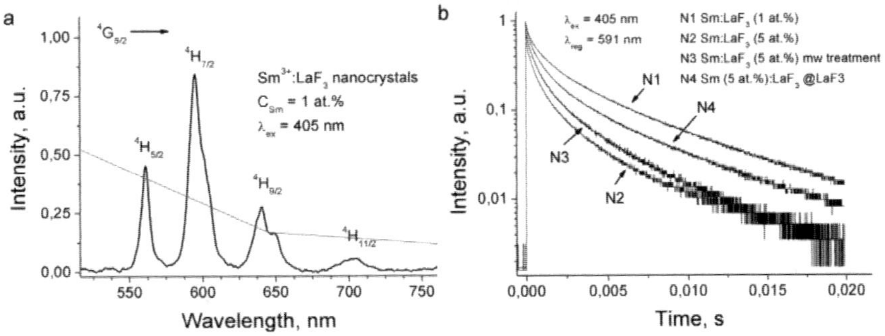

Fig.1. Room temperature luminescence spectrum (a) and luminescence decay curves (b) of LaF_3:Sm nanoparticles dried sample under 405 nm excitation.

Changes in preparation method have found reflection in luminescence decay of the samples. The decay curves of all four samples of Sm:LaF_3 nanoparticles are presented in Figure 1b. All samples of Sm:LaF_3 nanocrystals have exhibited non-exponential decay which is common for nanoparticles due to large surface to volume ratio and significant amount of surface perturbed sites [5] and which empirically can be divided into fast and slow components. The results of two-exponential fitting are presented in Table 1.

Table 1. Luminescence decay components of Sm:LaF_3 nanoparticles samples.

	N1 (C_{Sm} 1%)	N2 (C_{Sm} 5%)	N3 (C_{Sm} 5% microwaved)	N4 (C_{Sm} 5% core-shell)
t1, ms	4.79±0.02	3.27±0.02	3.33±0.02	4.26±0.02
t2, ms	0.85±0.02	0.67±0.02	0.66±0.02	0.77±0.02

The decay of heavier doped sample N2 is faster both in slow and fast components as it should be due to concentration quenching. Existence of LaF_3 shell lowers the decay times back apparently due to covering impurity sites at the surface thus eliminating surface quenching factors. The microwave treatment of nanoparticles on the contrary has not introduced any effect in implemented conditions which can speak for deficient fluorination or exposition during reaction.

The work was supported by the subsidy of the Russian Government (agreement No.02.A03.21.0002) to support the Program of Competitive Growth of Kazan Federal University and for the subsidy allocated to Kazan Federal University for the state assignment in the sphere of scientific activities [3.1156.2017/4.6].

References

[1] K. Binnemans, Chemical Reviews **109**, 4283 (2009)

[2] Q.Y. Zhang, X.Y. Huang, Progress in Materials Science **55**, 353 (2010)

[3] F. Wang, Y. Zhang, X. Fanc, M. Wang, J. Mater. Chem. **16**, 1031 (2006)

[4] J. Shao, Z. Wang, R. Wang, Z. Liu, L. Wang, J. Guo, X. Li, M. Zhang, C. Zhang, H. Zheng, J. of Solid State Chem. **49**, 199 (2017)

[5] M. Runowski, S. Lis, Journal of Alloys and Compounds **661**, 182 (2016)

Peculiarities of luminescence decay of Ce:LaF₃ nanoparticles depending on conditions of hydrothermal treatment

Elena Lukinova*, *Eduard* Madirov, *Maxim* Pudovkin, *Daria* Koryakovtseva, *Stella* Korableva, *Alexey* Nizamutdinov, and *Vadim* Semashko

Kazan Federal University, Institute of Physics, 18 Kremlevskaja str., Kazan, 420008 Russia

Abstract. Luminescence decay properties of crystalline LaF_3 nanoparticles containing 12 % of Ce^{3+} ions collected at different stages of microwave hydrothermal treatment were studied. The microwave treatment has led to improvement of surface perturbed sites luminescence together with reducing PL decay time of regular sites apparently due to higher surface/volume ratio.

1 Introduction

Fluoride crystalline nanoparticles and LaF_3 particularly doped with RE ions exhibit prospective luminescent properties. Due to low-energy phonons these materials provide advantages in up-conversion and photosensitizing demanding applications [1,2] and open novel applications for quantum electronics and photonics [3].
The variation of synthesis conditions which leads changes in energy transfer processes between impurities and influences the luminescence quenching factors [2] opens the way to manage characteristics of materials. Aim of this work is to study luminescence decay properties of Ce^{3+} ions in LaF_3 nanoparticles after microwave hydrothermal treatment.

2 Experimental

The nanoparticles of LaF_3 doped with 12 % of Ce3+ ions were synthesized via co-precipitation method [3-5] and subsequent modification of prepared nanoparticles using the microwave-assisted hydrothermal treatment was performed [6]. Four samples collected at different stages of fabrication were investigated. Namely N1 sample was obtained right after precipitation procedure. The N2 sample was collected right after the pH was adjusted to 5-6 by adding 25% solution of ammonium hydrate and then stirred with heating for 2 hours. The remaining suspension was placed into the microwave oven for 30 minutes (sample N3) and 180 minutes (sample N4). All samples were washed with distillate water and separated from the suspension by centrifugation. The resulting product was dried at 60 °C for 12 h.

* Corresponding author: elena.v.lukinova@gmail.com

 DOI: 10.1051/epjconf/201716103013

3 Results and discussion

Luminescence (PL) spectrum of Ce:LaF_3 sample N1 registered under 266 nm excitation (see Fig. 1a) appeared to be characteristic for all samples and coinciding with literature [2].

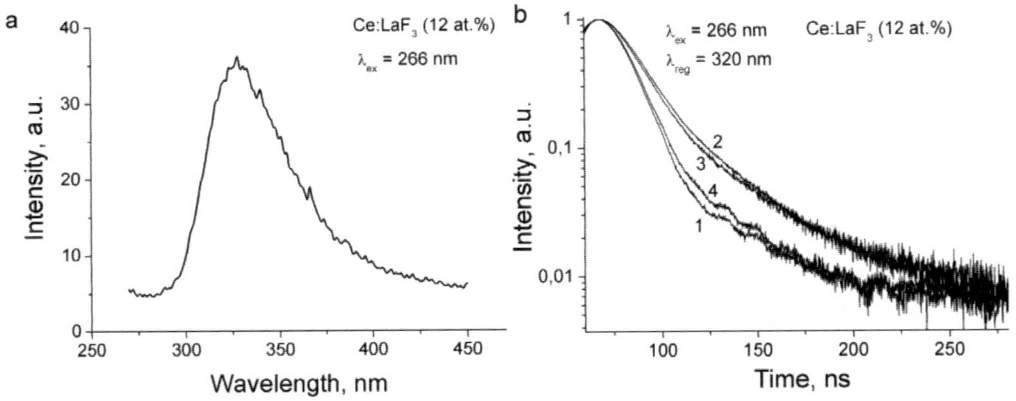

Fig. 1. Room temperature PL spectrum (a) and PL decay curves (b) of LaF_3:Ce (12 %) nanoparticles dried samples, 1 – N1 sample, 2 – N2 sample, 3 – N3 sample, 4 – N4 sample.

The decay curves of PL detected at 320 nm of all four samples of Ce:LaF_3 nanoparticles show fast and slow components (see Fig. 1b) determined by regular sites and perturbed sites relaxation respectively [2]. Results of the fitting the decays by two-exponential dependence are presented in Table 1. It is shown that both fast and slow components exhibit changes with synthesis procedure.

Table 1. Luminescence decay components of Ce:LaF_3 nanoparticles samples.

	LaF_3:Ce (N1)	LaF_3:Ce (N2)	LaF_3:Ce (N3)	LaF_3:Ce (N4)
t1, ns	55.6±0.5	36.6±0.5	38.2±0.5	43.1±0.5
t2, ns	8.1±0.5	9.8±0.5	10.1±0.5	7.9±0.5

It can be seen that stabilization of pH drops PL decay time for perturbed state and this can affect energy transfer to sensitizers. Also it is shown that microwave treatment somehow improves this effect together with reducing PL decay time of regular sites apparently due to higher surface/volume ratio.

The work was supported by the subsidy of the Russian Government (agreement No.02.A03.21.0002) to support the Program of Competitive Growth of Kazan Federal University and for the subsidy allocated to Kazan Federal University for the state assignment in the sphere of scientific activities [3.1156.2017/4.6].

References

[1] K. Binnemans, Chemical Reviews **109,** 4283 (2009)
[2] D. R. Cooper et al., Phys. Chem. Chem. Phys. **16,** 12441 (2014)
[3] M. S. Pudovkin et al., J. of Nanomaterials **2017,** 3108586 (2017)
[4] F. Wang, Y. Zhang, X. Fanc, M. Wang, J. Mater. Chem. **16,** 1031 (2006)
[5] J. Shao et al., J. of Solid State Chem. **49,** 199 (2017)
[6] E. M. Alakshin et al., J. of Nanomaterials **2016,** 7148307 (2016)

Ce^{3+} doped $LiYF_4$ nanoparticles fabrication by laser ablation

Elena Lukinova, *Ilnur* Farukhshin, *Alexey* Nizamutdinov*, *Eduard* Madirov, *Vadim* Semashko, and *Maxim* Pudovkin

Kazan Federal University, Institute of Physics, 18 Kremlevskaja str., Kazan, 420008 Russia

Abstract. Here we report on laser ablation approach to produce dielectric crystalline nanoparticles Ce:$LiYF_4$ inspired by method for metal nanoparticles fabrication in aqueous solutions. The obtained water solution has shown tangible presence of compound we believe to be agglomerates of Ce:$LiYF_4$ nanocrystals. The obtained powder has shown luminescence with peaks at 310 nm and 325 nm which is characteristic for the scheelite structured fluoride and luminescence decay quenching characteristic for Ce-doped nanoparticles.

1 Introduction

Fluoride crystalline materials are popular for luminescent applications due to their low-energy phonons and transparency in wide spectral range. These advantages are transferred also to nanomaterials applications e.i. upconversion and quantum cutting materials for bioimaging, sensibilization, phosphors and so on [1,2]. The compound of $LiYF_4$ is of specific interest due to isomorphic character of substitution for most of rare-earth ions (Yb, Tm, Gd, Eu, Tb, Er). There are works on propositions of $LiYF_4$ nanocrystals doped with Yb, Tm and Ce ions as efficient upconverters to UV spectral range [3]. One of the problems for $LiYF_4$ nanoparticles is complexity of synthesis procedure. Namely solvothermal method or thermal decomposition which is mostly implemented supposes high temperature exposure in autoclave or noble gas flow [1-3]. This could be costly when trying to transfer the method to large volumes of production.

Here we report on laser ablation approach to produce dielectric crystalline nanoparticles Ce:$LiYF_4$ inspired by method for metal nanoparticles fabrication [4].

2 Experiment details

The laser beam of 532 nm wavelength from Nd:YAG: laser was focused onto the surface of Ce:$LiYF_4$ crystal under the layer of distilled water as a target. The Nd:YAG laser operated in Q-switch regime with 10 ns pulses and 10 Hz repetition rate. Energy density of laser beam at the surface of the crystal was about 15 J/cm^2. After one hour of exposition the water from above the target crystal was collected. Part of the obtained sample was investigated by means

* Corresponding author: anizamutdinov@mail.ru

of dynamic light scattering technique (DLS), another part was dried and the acquired powder was tested by spectroscopic methods.

3 Results

The DLS investigation have shown presence of particles with the hydrodynamic radius of about 180 nm. The dried powder was luminescent under 289 nm excitation giving spectrum of $Ce:LiYF_4$ crystal. In the Figure 1A luminescence spectra of the $Ce:LiYF_4$ target and resulting powder are shown.

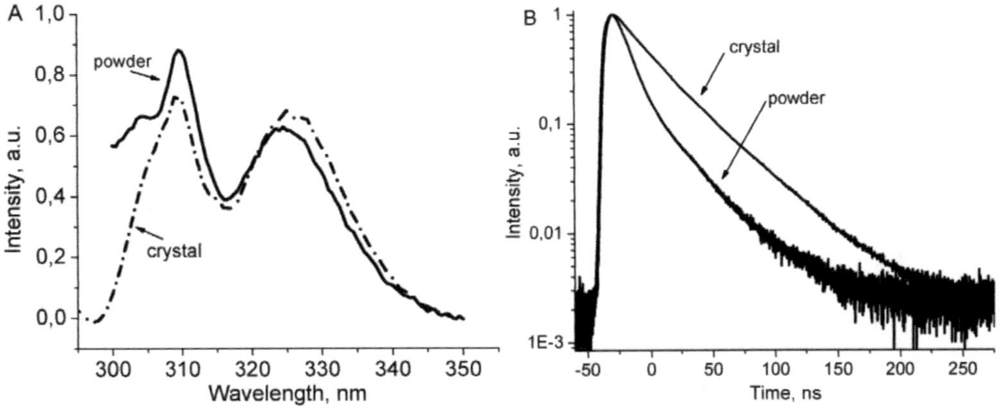

Fig. 1. A) $Ce:LiYF_4$ PL spectra of nanoparticles (powder) and monocrystal (crystal), B) luminescence decay for nanoparticles(powder) and monocrystal (crystal) detected at 311 nm.

The decay curve of obtained powder (Fig 1B) compared to bulk material reveals fast component usually attributed to surface quenching factors and slower component which is also quenched (21 ns vs 37 ns for bulk). The obtained luminescence decay picture is characteristic for nanosized crystalline material [5]. We believe that the hydrodynamic radius obtained is evidence of agglomeration of nanoparticles as it has been noticed in work [2]. Thus we propose the ablation of bulk $Ce:LiYF_4$ crystal in water produces crystalline nanoparticles of $Ce:LiYF_4$.

The work was supported by the subsidy of the Russian Government (agreement No.02.A03.21.0002) to support the Program of Competitive Growth of Kazan Federal University and for the subsidy allocated to Kazan Federal University for the state assignment in the sphere of scientific activities [3.1156.2017/4.6].

References

[1] S. Ye, R. Hu, N. Jiang, H. Wang, D. Wang, Dalton Trans. **44**, 15583 (2015)

[2] D. Zhang, G. De, L. Zi, Y. Xu, S. Liu, Mater. Res. Express **3**, 075005 (2016)

[3] J.A. Capobianco et al., Chem. Eur. J. **15**, 9660 (2009)

[4] F. Mafuné, J.-Y. Kohno, Y. Takeda, T. Kondow, H. Sawabe, Journal of Physical Chemistry B **39** , 9111 (2000)

[5] D.R. Cooper, K. Kudinov, P. Tyagi, C.K. Hill, S.E. Bradforth, J.L. Nadeau, Phys. Chem. Chem. Phys. **16**, 12441 (2014)

DOI: 10.1051/epjconf/201716103015

Quantum properties of a parametric four-wave mixing in a Raman type atomic system

A.V. Sharypov[1,2,*], Bing He[3], V.G. Arkhipkin[1], and S.A. Mislivets[1]

[1]Kirensky Institute of Physics, Federal Research Center KSC SB RAS, 50 Akademgorodok, Krasnoyarsk 660036, Russia
[2]Science Center "Newton Park", 1 Mira Str., Krasnoyarsk, 660049 Russia
[3]Department of Physics, University of Arkansas, Fayetteville, AR 72701, USA

Abstract. We present a study of the quantum properties of two light fields used to parametric four-wave mixing in a Raman type atomic system. The system realizes an effective Hamiltonian of beamsplitter type coupling between the light fields, which allows to control squeezing and amplitude distribution of the light fields, as well as realizing their entanglement. The scheme can be feasibly applied to engineer the quantum properties of two single-mode light fields in properly chosen input states.

Let us consider a three-level atomic system with the ground state |a> and two upper states |b> and |c>; see Fig. 1.

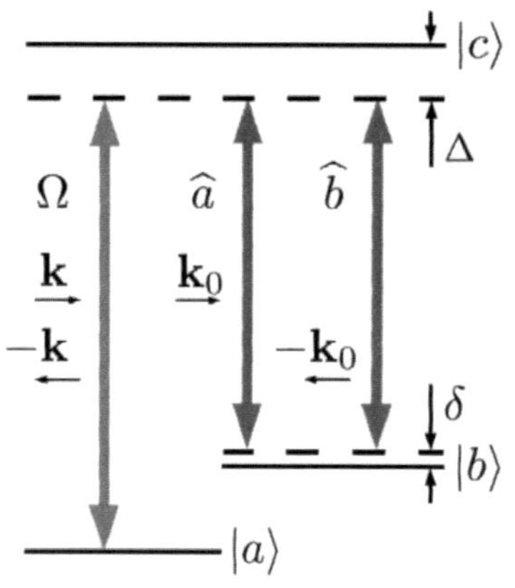

Fig. 1. Scheme of a Raman-type dispersive four-wave mixing.

Through the system the input quantum light fields of single mode, the blue ones in Fig. 1, effectively interact with one another. The classical standing electromagnetic field with the Rabi frequency Ω implements the transition |a>–|c>. Such classical standing wave is applied along z -direction and can be decomposed into two running wave with the wave vectors ±k.

* Corresponding author: asharypov@yandex.ru

DOI: 10.1051/epjconf/201716103015

The two counterpropagating quantum modes that are coupled to the transition |b>–|c> with the coupling parameter g also propagate along z-direction.

In order to avoid one- and two-photon transitions we use large one (Δ)- and two-photon (δ) detuning

$$|\,\Omega/\Delta| << 1, \qquad (1)$$

$$|\Omega\, g(n)\ 1/2/\,(\Delta\delta)| << 1, \qquad (2)$$

where g – is the coupling strenght.

The derived effective Hamiltonian for the present system has the similar form to a beamsplitter Hamiltonian. Further we demonstrate that there are possibilities for the parametric amplitude amplification (Fig.2) of a squeezed state, as well as the squeezing transferring (Fig.3), with such type of interaction.

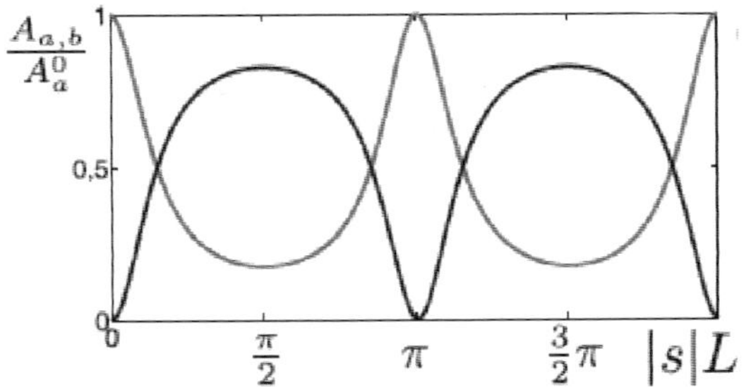

Fig. 2. Dynamics of the normalized quantum field's amplitude as a function of the dimensionless medium length.

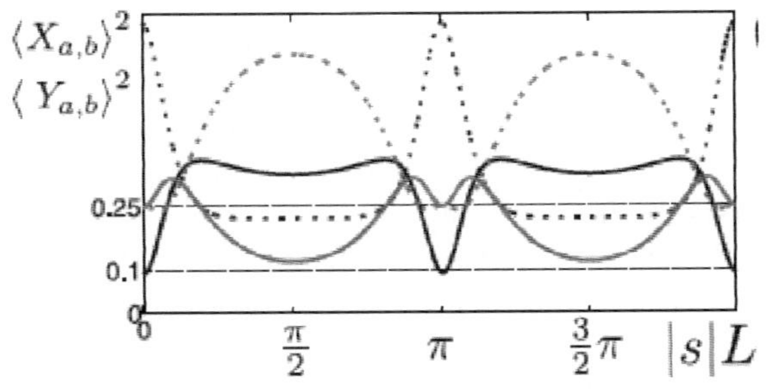

Fig. 3. Dynamics of the fields quadratures fluctuations as a function of the dimentionless medium length.

This work was supported by the Russian Foundation for Basic Research under Grant No 15-02-03959.

Spontaneous four-wave mixing in optical nanofibers at low temperatures

Anatoly Shukhin[1,*], and *Alexey* Kalachev[1,2]

[1]Kazan E.K. Zavoisky Physical-Technical Institute of the RAS, Kazan, 420029 Russia
[2]Kazan Federal University, Kazan, 420008 Russia

Abstract. In this work, spontaneous four-wave mixing in silica optical nanofibers is studied theoretically. The spectrum of generated biphoton field and its dependence on the nanofiber temperature are analysed.

1 Introduction

Currently, much attention is paid to developing sources of single photons and entangled photon pairs that are considered as basic ingredients for long-distance quantum communication, optical quantum computing, etc. [1]. One of the most promising approaches to the problem is the use of spontaneous parametric down-conversion and spontaneous four-wave mixing (SFWM) [2]. In particular, SFWM can be observed in widespread third-order nonlinear materials such as silica fibers. In order to increase the effective nonlinear efficiency, we can take advantage of optical micro/nano-fibers (MNF) [3]. However, it is well known that uncorrelated noise from spontaneous Raman scattering of the pump field can significantly decrease coincidence-to-accidental ratio [4]. To solve this problem, the nonlinear material can be cooled to liquid nitrogen [5] or even liquid helium [6] temperature. In the present work, we analyse spectral properties of SFWM in an optical MNF at such low temperatures.

2 Optical nanofibers

Fibers with a subwavelength waist diameter are fabricated from standard optical single-mode fibers using heating and stretching method [7]. By adjusting relative movement of fiber and flame during the preparation procedure, almost any adiabatic MNF profile can be made [8].

MNFs have a couple of features which make them pretty good for using as a nonlinear medium for the photon pair generation. First, perfect mode matching between tapered MNF and standard optical fiber can be achieved, which is important when the former is integrated in the telecommunication network. Second, due to the small nanofiber waist diameter, there is a large overlap between all interacting modes, so that efficiency of the SFWM process is high. This fact allows us to decrease the pump power and the length of the nonlinear medium, thereby making photon pair sources compact.

* Corresponding author: anatoly.shukhin@mail.ru

SFWM process is a nonlinear effect whereby two photons from the pump field are annihilated and simultaneously two photons in the signal and idler fields, which together called biphoton field, are created. As an example of MNF, we consider a tapered optical fiber with the waist diameter of 950 nm and waist length of 10 cm. Such transverse waist size is chosen to satisfy the phase matching condition leading to strongly nondegenerate SFWM, when the wavelength of the idler photons belongs to the transparency window of the silica fiber, while that of the signal photons belongs to the visible spectral range where detectors efficiency is high. Standard optical fiber, which MNF is assumed to be made of, has SiO_2 core of 10 μm diameter and Ge doped SiO_2 cladding of 125 μm diameter.

3 Joint spectral amplitude of the biphoton field

One of the most important characteristics of the biphoton field is the joint spectral amplitude (JSA) representing the probability amplitude of the photon pair creation as a function of the signal field and idler field wavelengths. In the present work, we study JSA as a function of temperature for MNF and analyse spectral properties of the biphoton field. Fig. 1 presents the results of theoretical simulation. It can be seen that the wavelength of the idler (signal) field decreases (increases) with the temperature, while spectral correlations between the emitted photons do not practically changed.

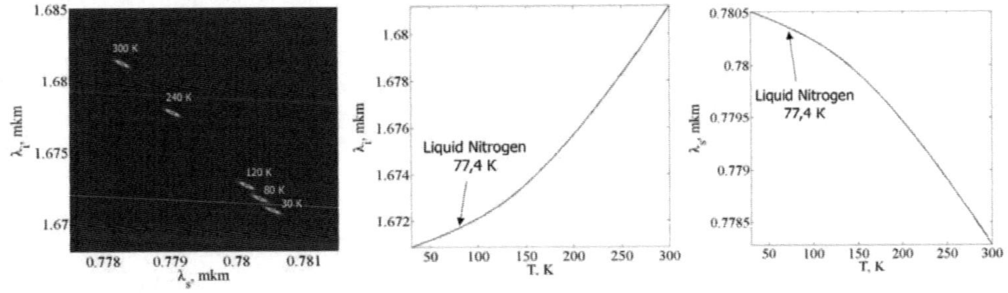

Fig. 1. Joint spectral amplitude of the biphoton field at different temperatures of MNF (left plot) and central wavelength of signal and idler photons as functions of the temperature. Pump wavelength is equal to 1064 nm, while the pump pulse duration is of order of 10 fs.

This work is supported by the Program "Topical Problems of Low-Temperature Physics" of the Presidium of the Russian Academy of Sciences.

References

[1] J. L. O'Brien, A. Furusawa, J. Vučković, Nature Photon. **3**, 687 (2009)
[2] M.D. Eisaman et al., Rev. Sci. Instrum. **82**, 071101 (2011)
[3] L. Cui et al., Opt. Lett. **38**, 5063 (2013)
[4] Q. Lin, F. Yaman, G.P. Agrawal, Phys. Rev. A **75**, 023803 (2007)
[5] H. Takesue, K. Inoue, Opt. Express **13**, 7832 (2005)
[6] S.D. Dyer, M.J. Stevens, B. Baek, S.W. Nam, Opt. Express **16**, 9966 (2008)
[7] L. Tong et al., Opt. Commun. **285**, 4641 (2012)
[8] J. M. Ward et al., Rev. Sci. Instrum. **85**, 111501 (2014)

EPJ Web of Conferences **161**, 03017 (2017) DOI: 10.1051/epjconf/201716103017
PECS-2017

Photophysical study of meso-phenothiazinyl-porphyrins metallocomplexes

Aleksander Starukhin[1,*], *Aleksander* Gorski[2], *Valery* Knyukshto[1], *Andrei* Panarin[1], *Tatiana* Pavich[1], *Luiza* Gaina[3], and *Emese* Gal[3]

[1]B.I. Stepanov Institute of Physics, National Academy of Sciences of Belarus, Minsk, Belarus
[2]Institute of Physical Chemistry, Polish Academy of Sciences, Warsaw, Poland
[3]Babes Bolyai University, Cluj-Napoca, Romania

Abstract. Photophysical parameters of a set of metallocomplexes of meso-phenylthiazinylporphyrins with Zn (II), Pd (II) and Cu (II) ions were studied in different organic solvents, solid solutions and polymeric matrices at room and liquid nitrogen temperatures. The dependence of the spectral and photophysical parameters on changing the molecular structure with increasing number of branched substituents attached to aryl groups in different positions of the porphyrin macrocycle has been established.

Macroheterocyclic compounds are widely used in medicine (compounds for photodynamic therapy and photodiagnostics) and as materials for design of electronics components (sensors, light emitting systems, solar cells and etc.), as well as systems for creation of optical memory, optical switches and etc. At the same time, increasing number of new applications of organic semiconductors in molecular electronics requires synthesis and studding of new molecular systems and optimization of their parameters.

The metallocomplexes of meso-phenylthiazinylporphyrins with Zn (II), Pd (II) and Cu (II) ions were studied in the current work. The synthesis of compounds was carried out by modified methods described in [1]. The structures of investigated compounds are shown on Fig. 1. The photophysical characteristics for newly synthesized metallocomplexes in various organic solvents and polymer films are presented (Tab. 1).

The progressive substitution with additional phenylthiazinyl (PTH) groups in the meso-positions of the porphyrin macrocycle leads to rather small but noticeable changes in the porphyrin absorption and fluorescence spectra (Fig. 2). It is well known [2] that with increasing of number of branched substituents in the meso-position of the porphyrin macrocycle the energy of the a_{2u} MO is changing, so that $^1E(a_{2u}, e_g) < {}^1E(a_{1u}, e_g)$. The substitution with PTH fragments in the meso-positions of the porphyrin macrocycle leads to the changes in the photophysical properties of the studied compounds. For example, in going from the parent unsubstituted Zn-TPP to the derivatives with increasing number of the PTH groups (Zn-TP2A, Zn-TP2B; Zn-TPy4PtzA, Zn-TP2D) the number of photophysical parameters was found to be altered (Fig. 2 and Tab. 1).

* Corresponding author: a.starukhin@ifanbel.bas-net.by

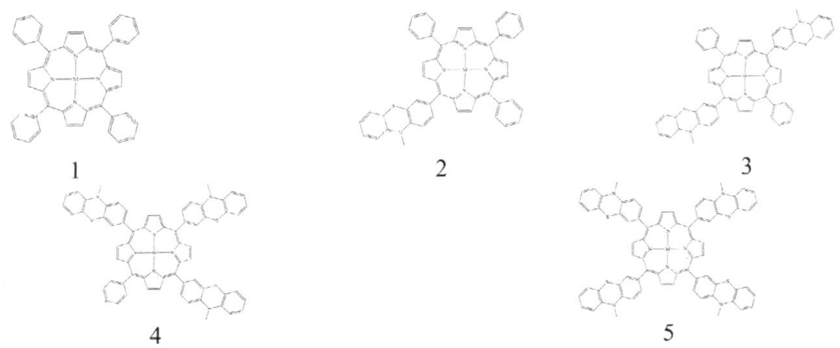

Fig. 1. Structure of meso-phenylthiazinylporphyrins metallocomplexes (where M- Zn(II), Pd(II) and Cu(II)),: 1-M-5,10,15,20-tetraphenylporphyrin (M-TPP); 2- M-TP2A; 3-M-TP2B; 4- M-TPy4PtzA; 5 -M-TP2D.

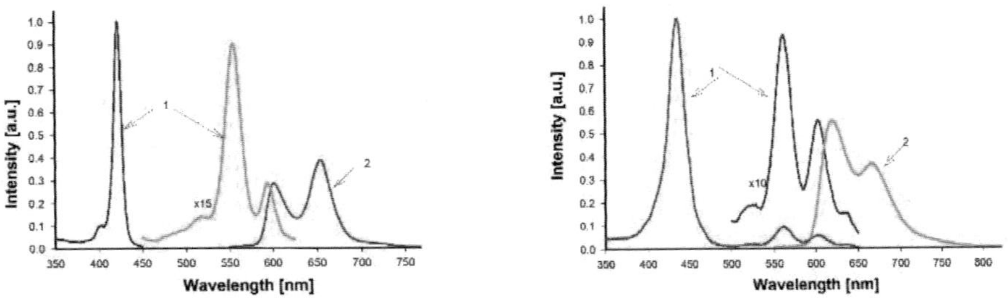

a b

Fig. 2. Absorption (1) and fluorescence (2) spectra of Zn-TPP (a) and Zn-TP2D (b) in 2-m-tetrahydrofurane at 300 K.

Table 1. Photophysical parameters of Zn-meso-phenylthiazinylporphyrins in 2-m-tetrahydrofurane at 300 K.

Compound	λ_{max} Sore, nm	$\lambda_{0\text{-}0}$, nm	Fluorescence $\lambda_{0\text{-}0}$,nm	φ_{Fl}	τ_{Fl}, ns
Zn-TPP	422	594	601	0,028	1,82
Zn-TP2A	427	599	608	0,033	1,24
Zn-TP2B	428	599	611	0,034	1,26
Zn-TPy4tzA	430	602	615	0,03	1,10
Zn-TPP2D	435	602	619	0,025	1,02

First of all, the fluorescence quantum yield φ_{Fl} increases with increasing of number of attached PTH groups. A measurement of the fluorescence decay times demonstrates that decay becomes, as rule, also faster.

This work was supported by the BRFFI Grant No. F16RA-007 and H2020-MSCA-RISE-2014-RISE-645628 - METCOPH project.

References

[1] B. Brem, Em. Gal, L. Gaina et al., Dyes and Pigments **123**, 12 (2015)

[2] M. Gouterman, J. Mol. Spectr. **6**, 138 (1961)

Generation of single-photon states with orbital angular momentum in the process of spontaneous parametric down-conversion in a cavity

Dinislam Turyikhanov[1,*], *Andrey* Shkalikov[1], *Ilnur* Latypov[1], and *Aleksey* Kalachev[1,2]

[1]Kazan E. K. Zavoisky Physical -Technical Institute, 10/7 Sibirsky tract str., Kazan, 420029 Russia
[2]Kazan Federal University, 18 Kremlyovskaya str., Kazan, 420008 Russia

Abstract. We consider the generation of single-photon states with orbital angular momentum in the process of spontaneous parametric down-conversion in a nonlinear crystal placed in a cavity. For creating optical beams with an orbital angular momentum, a ring resonator with a Dove prism is proposed to be used.

Light beams with orbital angular momentum (OAM) are currently the subject of active research. In contrast to the spin angular momentum, which is associated with the polarization of light, OAM arises due to the spatial distribution of the intensity and phase of the optical field. Promising applications of OAM are associated with manipulation of microparticles [1], increasing the capacity of optical communication channels [2], increasing the stability of quantum communication through the atmosphere [3] and a number of other applications.

One of the simplest ways of generating single photons is the process of spontaneous parametric down-conversion (SPDC) of light, during which a photon from the pump beam incident on a nonlinear crystal is annihilated with creating two photons called idler and signal. This process satisfies the energy and phase matching conditions: $\boldsymbol{\omega_p} = \boldsymbol{\omega_i} + \boldsymbol{\omega_s}$ and $\vec{k}_p = \vec{k}_i + \vec{k}_s$, where ω и $\vec{k}$ are the frequency and wave vector of signal (i), idler (s) and pump (p) photons. In addition, for collinear SPDC we have the following conservation condition [4]:

$$\ell_{pump} = \ell_s + \ell_i \,, \tag{1}$$

where ℓ_{pump}, ℓ_s and ℓ_i are the values of OAM for the pump, signal and idler photons, respectively. The sum of OAM values of the created photons is equal to OAM value of the pump. This is confirmed by coincidence measurements with detecting photons of the signal and idler SPDC fields, when the crystal is pumped by Laguerre-Gaussian beams with different OAM. The probability of coincidence is calculated by the formula $P_{p_s,p_i}^{\ell_s,\ell_i} = |C_{p_s,p_i}^{\ell_s,\ell_i}|^2$, where the probability amplitude $C_{p_s,p_i}^{\ell_s,\ell_i}$ is given by the integral [5]

$$C_{p_s,p_i}^{\ell_s,\ell_i} \sim \int dr_\perp \, \Phi(r_\perp) \left[LG_{p_s}^{\ell_s}(r_\perp)\right]^* \left[LG_{p_i}^{\ell_i}(r_\perp)\right]^*, \tag{2}$$

* Corresponding author: 7intur@gmail.com

where $r_\perp$ is the radial coordinate in the transverse X–Y plane, $\Phi(r_\perp)$ is the distribution of the amplitude of the pump field, $LG_{p_s}^{\ell_s}(r_\perp)$ and $LG_{p_i}^{\ell_i}(r_\perp)$ are the distributions of the amplitude of the signal and idler fields, respectively. Figure 1 shows the theoretical simulation of the coincidence probability.

In this paper, we consider the generation and detection of single photons with OAM via the SPDC process in a nonlinear crystal placed in a cavity. A cw neodymium laser at 532 nm is assumed to be used as a pump, while a PPLN crystal in which correlated photon pairs are produced at 810 nm and 1550 nm can be used as a nonlinear material. The nonlinear crystal is placed in a cavity whose mirrors have a high reflection coefficient for one of the emitted fields, say signal, and are transparent for the second emitted field and pump field. Beams of the signal field with OAM can be created in a ring cavity consisting of one planar and two spherical mirrors [6]. Such geometry of the resonator in the form of an obtuse triangle allows one to reduce the influence of astigmatism of the spherical mirrors. Rotation of the field around the beam axis is realized by means of a Dove prism. If the cavity creates specific value of OAM for the signal field, the corresponding OAM value for the idler field satisfying the conservation law (1) should be observed. When the cavity field proves to be a superposition of Laguerre-Gaussian modes, the corresponding superposition should be observed for the conjugated field.

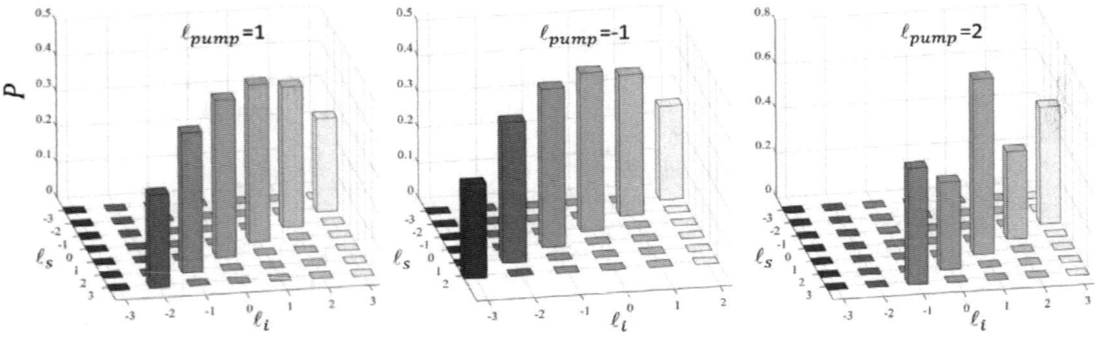

Fig. 1. The coincidence probability P for the orbital angular momenta of the signal and idler photons for ℓ_p= 1, –1, and 2. The result of simulation using Eq. (2).

This work was supported by the Russian Foundation for Basic Research (grant № 16-02-01055).

References

[1] M. E. Friese, J. Enger, H. Rubinsztein-Dunlop, and N. R. Heckenberg, Phys. Rev. A **54** 1593 (1996)

[2] G. Gibson, J. Courtial, M. Padgett, M. Vasnetsov, V. Pas'ko, S. Barnett, and S. Franke-Arnold, Opt. Express **12**, 5448 (2004)

[3] M. Malik, M.O'Sullivan, B. Rodenburg, M. Mirhosseini, J. Leach, M.P.J. Lavery, M.J. Padgett, and R.W. Boyd, Opt. Express **20**, 13195 (2012)

[4] S. Walborn, A. De Oliveira, R. Thebaldi and C. Monken, Phys. Rev. A **69** 23811 (2004)

[5] S. Franke-Arnold, S.M. Barnett, M.J. Padgett, L. Allen, Phys. Rev. A **65**, 033823 (2002)

[6] E. Abramochkin, N. Losevsky, V. Volostnikov, Opt. Commun. **141**, 59 (1997)

Stationary and coherent spectroscopy of $^{167}Er^{3+}$ in waveguides in $^{7}LiYF_4$ crystal

Mansur Minnegaliev[1], *Ravil* Urmancheev[1,*], *Konstantin* Gerasimov[1], *Alexander* Kalinkin[2], *Ivan* Dyakonov[2], *Mikhail* Saygin[2], *Sergey* Kulik[2], *and Sergey* Moiseev[1]

[1]Kazan Quantum Center, Kazan National Research Technical University n.a. A.N. Tupolev-KAI, 18a Chetaeva Str., Kazan, 420126 Russia
[2]M.V. Lomonosov Moscow State University, Faculty of Physics, Leninskie Gory, Moscow, 119991 Russia.

Abstract. We have conducted a spectroscopic investigation of $^{167}Er^{3+}$ ions in optical waveguides on an optical transition between the hyperfine sublevels of $^4I_{15/2}$ and $^4I_{9/2}$ multiplets. Waveguides with diameters ranging from 20 to 100 μm were produced in the crystal by a femtosecond laser using the depressed-cladding approach. The spectroscopy results of $^{167}Er^{3+}$ ions inside the waveguides show additional broadening and an overall shifts of the spectra compared to the bulk spectrum of ions. The sign of the observed frequency shift depends on the diameter of the specific waveguide. We have also observed a two-pulse photon echo in several waveguides. The acquired results show the possibility for integrated quantum schemes in rare-earth ions doped crystals.

Development of optical quantum technologies including optical processors and quantum repeaters is accompanied by the increased attention to their subsequent realization in optical integral schemes. That means that experimental development of integral quantum memories in such schemes is also necessary. Multi-atomic systems can serve as active media for such experiments, such systems include crystals that often have longer quantum coherence times. So waveguides in crystals doped with rare-earth ions are of great interest especially those combinations of rare-earth ion and crystal matrix that exhibit desirable spectroscopic parameters for quantum memory realization. In our previous work we studied the $^{167}Er^{3+}$:$^{7}LiYF_4$ crystal [1], where we observed a 90 MHz inhomogeneous linewidth for the optical $^4I_{15/2}(\Gamma_{56})$ – $^4I_{9/2}(\Gamma_{78})$ transition. Then, thanks to the advances in the crystal growth techniques this value was lowered down to 24 MHz [2]. Such narrow inhomogeneous linewidth of optical transition makes this crystal a promising candidate for Raman quantum memory realization.

In this work we present the results of spectroscopic study of $^{167}Er^{3+}$ ions in symmetrical waveguides in $^{7}LiYF_4$: $^{167}Er^{3+}$ crystal. Five 15.8 mm long waveguides with diameters of 20, 30, 50, 75 and 100 μm and were produced by a femtosecond laser using the depressed-cladding approach with 515 nm wavelength. We carried out a high-resolution spectroscopy of $^{167}Er^{3+}$ ions in different waveguides at 4 K temperature. Fig. 1a shows three of the resulting

* Corresponding author: ravil@kazanqc.org

spectra in π polarization, for 30, 100 μm waveguides and for the bulk of the crystal. The absorption lines of ions inside the waveguide are broadened due to additional inhomogeneities introduced by the local tensions caused by the waveguide creation process. The optical density falls subsequently since the atomic absorption lines are spread ever the wider area. Fig. 1a also shows that the spectra obtained in the waveguides are also shifted from the central frequency defined by the bulk spectrum. For 100, 75 μm waveguides $^{167}Er^{3+}$ spectra are shifted to the lower frequency region, for 50 μm waveguide the spectrum (not shown on the figure) stays in the center and 30 and 20 μm waveguide spectra experience the opposite sign shift to higher frequencies.

We have also investigated the system by coherent spectroscopy techniques. It is worth noting that despite the absorption lines widening we managed to observe two-pulse photon echo in some of the waveguides (see fig. 1b). At the same time for the crystal studied in [1] with 90 MHz inhomogeneous linewidths we could not see the echo. The photon echo experiments were conducted on the single absorption line that is located around the zero frequency on fig. 1b. The first obtained estimations reveal that the transverse relaxation time for this transition is at least 500 ns. The obtained results show that quantum integrated schemes are possible to realize in the studied $^{7}LiYF_4$: $^{167}Er^{3+}$ crystal.

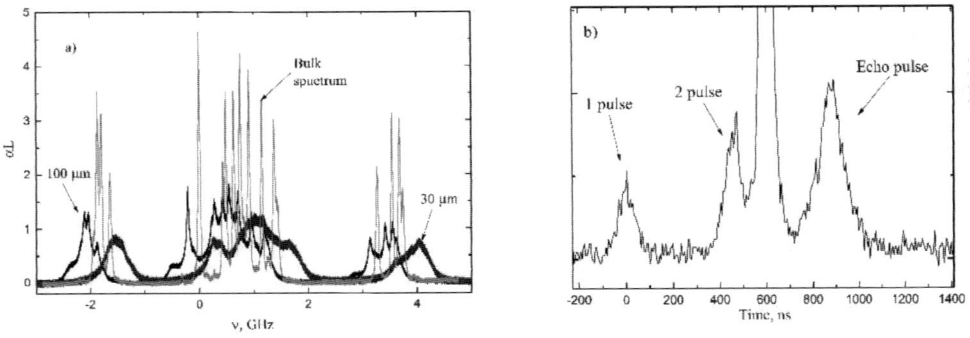

Fig. 1. a) $^{167}Er^{3+}$ ions absorption spectra in $^{7}LiYF_4$ crystal in bulk (light gray curve), in 100 μm diameter waveguide and in 30 μm diameter waveguide (two black curves). Zero on the frequency scale corresponds to the ν=370,574 THz. b) Two-pulse photon echo signal for 100 μm waveguide. The delay between the first two pulses equals 450 ns.

This work is supported by Russian Foundation for Basic Research grant #17-02-00918.

References

[1] K.I. Gerasimov, M.M. Minnegaliev, B.Z. Malkin, E.I. Baibekov, S.A. Moiseev, Phys. Rev. B **94**, 054429 (2016)

[2] M.M. Minnegaliev, E.I. Baibekov, K.I. Gerasimov, S.A. Moiseev, M.A. Smirnov, R.V. Urmancheev, Quant. Electron. **47**(9), 778 (2017)

AUTHOR INDEX

AUTHOR INDEX

EDP Sciences – Web of Conferences
17, Avenue du Hoggar
Parc d'Activité de Courtabouf
BP 112
F-91944 Les Ulis Cedex A
France

CURRAN ASSOCIATES INC.
proceedings.com

ISBN 978-1-5108-5103-0